Ethnicity in the Fiction of Lummer Dai and Yeshe Dorjee Thongchi

A New Historicist Approach

Dr. Juri Dutta

ADHYAYAN

ADHYAYAN PUBLISHERS & DISTRIBUTORS
New Delhi-110002

Published by
ADHYAYAN PUBLISHERS & DISTRIBUTORS
4378/4B, 105, J.M.D. House, Murari Lal Street
Ansari Road Darya Ganj, New Delhi - 110002
Ph.: 011-23263018, 011-23277156, Fax: 23280028
Email: adhyayanpublishers@yahoo.com

Ethnicity in the Fiction of Lummer Dai and Yeshe Dorjee Thongchi A New Historicist Approach

Edition 2012
ISBN 978-81-8435-322-8

Printed in India

Published by Harish Chandra Yadav for Adhyayan Publishers & Distributors, Delhi. Laser Typesetting at Jain Computers and Printed at Tarun Offset Printers. Delhi

Dedicated to My Father
Late Lakshinath Dutta
My Mother
Smt. Kunjalata Dutta
and
my Mother-in-law
Late Manorama Devi

Acknowledegments

My sincere thanks go to Dr. Sucheta Sen Chaudhuri, who guided me for my Ph.D. work. I thank Dr. Sarit Kumar Chaudhuri for his constant reminder to publish this book. I am grateful to my family members who have been helpful in encouraging me with their material and psychological support. My helping hands deserve special thanks for relieving me of the household choirs. Prince has always been there by my side as a moral and intellectual support. It would not have been possible for me to get the book published without his interest and guidance. I am thankful to Prince and my son Nunu.

Juri Dutta

Contents

Introduction

Lummer Dai (1940) and Yeshe Dorjee Thongchi (1952) are two pioneering litterateurs of Arunachal Pradesh. Both of them happen to be first generation literates who rose to important bureaucratic positions in the state. While Lummer Dai was serving as the Director, Department of Information and Public Relation when he passed away in 2002, Yeshe Dorjee Thongchi presently holds the post of Commissioner to the Government of Arunachal Pradesh. Quite obviously therefore, both of them not only had access to governmental policy making but were also abreast of the transformations taking place in the urban Arunachal of today. However, as both of them are first generation literates, coming originally from rural Arunachal, they also have a sense of affiliation with their root. This obviously provides them with a two-fold advantage - they can keep shifting between two worlds with which they are equally familiar and this probably makes them the best-equipped people to write about the Arunachal Pradesh going through a process of transition.

Dai and Thongchi belong to tribal communities of Arunachal Pradesh and they could not receive education in their respective mother tongues and therefore, they wrote in Assamese, the language that was the medium of instruction in educational institutes of Arunachal Pradesh till 1971 (Chowdhury, 1983: 315). In fact, both of them have cited different reasons for the selection of the

Assamese language for their writings. Kaling Borang, Lummer Dai's nephew, says:

> Probably because of the fact that the North East Frontier Agency and Assam were adjacent to each other, the selectors found it easier for selecting teachers from Assam. These Assamese teachers were the pioneers who taught in schools even in far-flung areas ... They felt [at] home in NEFA as it was a part of the once undivided Assam ... They were devoted and dedicated to their duties with heart and soul ... Therefore, the pioneering teachers are still remembered by many senior officials with gratitude and reverence. Lummer Dai is one example of students taught by those pioneering dedicated Assamese teachers (2006: 4).

As Assamese was the medium of instruction right from the level of primary school, it was quite close to being the mother tongue of the Arunachalese. In an interview with Ajanta Dutta, Lummer Dai says, "I think that mother tongue is the best medium to express one's feelings. Assamese is not my mother tongue. But I have an association with the Assamese language since the time I was being suckled by my mother" (*Aji*, 3rd February 2002).

Similarly Yeshe Dorjee Thongchi has said that the chief reason behind his choice of the Assamese language for his writings was the fact that he was taught in Assamese medium from the days of primary school. In an interview with Girin Tamuli and R. N. Koley, he says:

> ... I had a personal liking for the [Assamese] language. From a practical point of view, Assamese is the language which is accessible to the majority of our people. People coming down from the hills to the plains interact, exchange views and feelings through the Assamese language (2006: 44)

There are just a handful of writers from Arunachal who write in Assamese. Though Assamese was the medium of instruction throughout the schools of Arunachal Pradesh till the early eighties, the new generation is distinctly separated from Assamese language after the introduction of Hindi. Earlier Assamese was used as the link language among different tribes as very often the language of one tribe is not comprehensible to another. Assamese was widely used in meetings, public gatherings and market places by different tribal communities of the state as a lingua franca. At present, people of the foot hills are still well conversant in Assamese. Lummer Dai talked about the feeling of brotherhood between people of the states of Assam and Arunachal in the annual meeting of Lakhimpur District Sahitya Sabha in 2002:

> The Arunachalee people understand the Assamese mind. And the Assamese people know the heart of the Arunachalese. There is a brotherly relation between the Arunachalese and the Assamese. We are brothers and sisters (Udayaditya Bharali: *Asomiya Pratidin,* 10th April 2002).

Both Dai and Thongchi started their writing careers when they were still in primary school. Lummer Dai started his career with an article 'Aboror Rup Nuguse Kiyo?' (Why don't the Adis Change?) for the hand-written school magazine *Giribani* when he was in his fourth standard. Thereafter, there was no looking back for Dai as he continued to write articles, short stories and novels prolifically in different magazines. His published novels are *Paharar Xile Xile* (In the Midst of the Hills), *Prithivir Hanhi* (The Smile of the Earth), *Mon aru Mon* (Heart to Heart), *Kanyar Mulya* (Bride Price) and *Upar Mahal* (The Higher Level) in 1961, 1963, 1965-66, 1978 and 2002 respectively.

Likewise, Yeshe Dorjee Thongchi also accepts the fact that the "zeal for creative writing was rooted in my heart in early school days, probably, when I was a student of class V or VI" (2006: 42). Thongchi started his career as a poet; he contributed a poem titled '*Jonbai*' for the children's magazine of the same name. Next he took to writing plays. Finally he switched over to short stories and novels and it was the narrative form "which eventually accommodated" (ibid.) him. Thongchi's published novels are *Sonam* (1981), *Lingjhik* (1983), *Mouno Ounth Mukhar Hriday* (Silent Lips Murmuring Hearts), (2001), and *Saba Kata Manuh* (The Man Who Slices Dead Bodies) (2004).

Lummer Dai and Yeshe Dorjee Thongchi introduce the readers to the life and society of Arunachal Pradesh in all its variety. The ethnic, cultural and traditional varieties of the state are best reflected in the literary works of these writers. In their writings many undiscovered aspects of tribal life and society of Arunachal Pradesh are beautifully reflected. The variety, the undiscovered wealth of the cultural, traditional and ethnic aspects of different tribes are not yet analyzed or studied through literature in Arunachal Pradesh. For a pseudo-anthropologist or for a curious ethnographer it is possible to look at the tribal culture like an "alien gazer" who might see the "strange and exotic aspect" of it (Mishra, 2001: 33). But a person, who is fortunate to be both an insider as well as a creative writer, can write with the perspective of one who constitutes an integral part of that very culture. Therefore, an attempt is made in this book to look at the life and society of Arunachal Pradesh as represented in the literature of Dai and Thongchi. In fact, Dai and Thongchi's novels may be considered to be "ethnographic" for its vivid representation of the festivals, ways of worship, rituals, and marriage systems of different tribes of Arunachal. However, in doing this, I am aware of the fact that to treat the relationship between fiction and life as direct and unproblematic would be a fallacy. My study will be informed by an awareness of the pitfalls of a "naïve ethnographic or anthropological reading" which treats fiction as though it transparently and directly represents the world of a culture, ignoring the politics of representation and of subjectivity (Snyder, 2010: 179). As such, while I value the importance and the social value of such a study, I am aware of the complexity involving the relationship between a work of art and the writer.

While analyzing the novels of Lummer Dai and Yeshe Dorjee Thongchi, I have some objectives in mind. I believe that to understand and appreciate the fictions of these two writers, one needs a minimum level of familiarity with the tribal life and society of Arunachal Pradesh. Therefore, I will begin with a description of the tribal communities of the state as reflected in the writings of social scientists in the first chapter. Arunachal Pradesh is the homeland of twenty-six major tribes and more than hundred sub-tribes. These varieties of tribes with their customs and traditions are discussed here.

Following this, I shall discuss the thematic concerns of Lummer Dai and Yeshe Dorjee Thongchi's novels in the second chapter. Lummer Dai, an Adi, portrays the Adi social life with its rituals, beliefs and customs in his novels. At the same time, the novels represent the social transformation taking place in Arunachalee society. Similarly, Thongchi, a Sherdukpen, depicts the social customs of the Sherdukpen and the Monpa tribes in his novels. Almost all his novels are set against one or more traditional customs of the Sherdukpen or the Monpa tribes of the state. His novels reflect an attempt to effect unity in diversity amongst the tribes of Arunachal. This chapter discusses these aspects of the novels of the two writers.

The third chapter will provide a discussion of the literary and cultural theory of New Historicism and the way in which it can be applied to a study of fiction. New Historicism attends primarily to the historical and cultural conditions of the production of a text in place of dealing with a text in isolation. Novels are products of a particular author, of a particular time, and of a particular place. And therefore, novels can be best analyzed with the help of a theory that takes a holistic look at the background and ideology of a literary work. As New Historicism is a theory that can address these very issues, it has been used here to analyze the novels of Lummer Dai and Yeshe Dorjee Thongchi.

Chapter IV studies the novels of Lummer Dai in the light of New Historicism. Dai's novels portray the cultural life of his own (Adi) community in a realistic manner. But along with this realistic description the beliefs and ideologies of a whole society percolate down to these novels, sometimes, in spite of the writer himself.

Chapter V applies the New Historicist approach to study the novels of Yeshe Dorjee Thongchi. Thongchi's novels transport us to the world of the Sherdukpen and Monpa societies. However, like Dai, Thongchi's novels are not just windows to these worlds but are also bearers of the archetypal ethos and world-views by virtue of which a society defines itself and also asserts its distinctiveness.

Methodology

The proposal for methodology draws on the New Historicist assumptions, the principal one being the need to firmly place the text in social practice. New Historicists try to establish a relation between literature and the culture of a period. It tries to see the relation of the literary work with the time and place to which it belongs. New Historicism concerns itself with the collective beliefs, social practices and cultural discourses that shape a particular work. In any case, it is imperative to situate Lummer Dai and Yeshe Dorjee Thongchi's novels in the period, culture and society to which it belongs. I interviewed Ms. Nani Dai (the wife of late Lummer Dai) and Yeshe Dorjee Thongchi to be familiarized with the social milieu from which they come. I have reviewed articles written on them and quoted from interviews conducted by me and others on these two writers. Ethnographic literature is consulted and referred to for discussing tribal life and culture of Arunachal Pradesh.

1

Communities of Arunachal Pradesh

Section A

The Chambers 21st Century Dictionary defines the word 'tribe' as "a group of people, families, clans or communities who share social, economic, political, etc ties and often a common ancestor and who usually have a common culture, dialect and leader" (1997: 1504). According to *The New Millennium Pronouncing Anglo-Assamese Dictionary* 'tribe' is a large family that has descended from a common ancestor. Tribe is here defined as a primitive race or a group of related families governed by a chief (2002: 843). It is possible to define the word 'tribe' differently. The concept of tribe can not be homogenous throughout the world as there is dissimilarity of opinion amongst scholars. However, although a comprehensive definition of tribe is not possible to set forth, it is possible to find out the characteristics of a tribe analyzing different definitions by various scholars. Kinship ties, common territory, one language, joint ownership, one political organization, absence of internecine strife have all been referred to as the main characteristics of a tribe (Majumdar and Madan,1988: 240). But some anthropologists have not accepted the above characteristics of a tribe. According to Rivers (quoted in ibid. 1988), habitation in a common territory is not a vital feature of tribal organization. But others like Perry (quoted in ibid. 1988) have insisted on it saying that even nomadic tribes roam about within a definite region.

Majumdar (1988) defined tribe as a group with territorial affiliation, practicing endogamy, with no specialization of functions, ruled by tribal officers, hereditary or otherwise, united in language or dialect, recognizing social distance with other tribes, or castes, and above all, conscious of a homogeneity of ethnic and territorial integration. In his essay "Transformation of Tribes and Analogous Social Formations", B.K. Roy Burman says:

> The term (tribe) refers to a stage of social formation in an evolutionary scheme of development of technology, knowledge of and capacity for control of the forces of nature, method of transmission of the same, perception of man's relation with man and with nature and scale of organization of social groups. (1992:28).

According to Morgan "Each tribe was individualized by a name, by a separate dialect, by a supreme government, and by the possession of a territory which it occupied and defined as its own" (1996: 103). Though Morgan's views on society and social organization have not been universally accepted, one has to admit that Morgan provides a strong base for social anthropologists. In *Ancient Society* (1877/1996), he has formulated a few attributes of an Indian tribe which he has discussed under several propositions.

(i) The possession of a territory and a name
(ii) The exclusive possession of a dialect
(iii) The right to invest sachems and chiefs elected by the gentes

(iv) The right to depose these sachems and chiefs
(v) The possession of a religious faith and worship
(vi) A supreme government consisting of a council of chiefs
(vii) A head-chief of the tribe in some instances.

It has been already said that anthropologists, historians, political theorists, sociologists, and others are yet to arrive at a clear and general definition of tribe having wide acceptance. The disagreement is larger today than it was in the past due to their study of greater variety of societies in the present. In his book *Society and Politics in India: Essays in a Comparative Perspective*, Andre Beteille (1999) says that in the study of Indian tribes, the anthropologists do not confine themselves to the present or even to the recent past, but try to go back to medieval times. He has made a distinction between two approaches, that is, - evolutionary and historical. The evolutionist approach takes a long range view of the passage of time and stresses the succession of social formations. The historical approach limits itself to a particular framework of space and time and stresses the coexistence of different social formations within that framework. As such, the historical perspective is a much broader area than the evolutionist approach. Therefore, the presence of survival, which is a necessary component in the historical framework, is regarded as an anachronism in the evolutionist perspective (Beteille, 1999: 58).

Morgan's concept of tribe was also rooted in the evolutionary perspective. Godelier (ibid. 1999: 61) has also accepted the need to define tribe in an evolutionary perspective. Among recent attempts to provide a general definition of tribe, Marshall Sahlins has received some attention. According to Sahlins,

> A tribe is a segmental organization. It is composed of a number of equivalent, unspecialized, multifamily groups, the structural duplicate of the

> other: a tribe is a congeries of equal kin group blocs (quoted in ibid. 1999: 60).

Though the evolutionary perspective has been accepted and adopted by many anthropologists, it is not that all anthropologists who write about tribes follow it. Morton Fried (ibid. 1999: 62) has put forward a different notion of tribe from Morgan and Godelier. He argues that the definition and concept of tribe used by anthropologists is inherently ambiguous. Morgan's definition of tribe which Godelier repeats is that a tribe is a totally organized society. When we speak of a tribe as a completely organized society, we assume that there are boundaries separating tribes from each other. However, it is not clear what the nature of these boundaries is as there is no such clear boundary between one tribe and another as that between tribe and non-tribe.

Fried (ibid. 1999: 64) also puts forward his argument against the opinion that each tribe has its own distinctive language which defines its boundaries. There are well known examples of people speaking the same language, being divided into several endogamous groups as well as examples of people intermarrying though their native languages are different. Moreover, some of the tribes have no separate language of their own but use the language prevalent in the region they inhabit, particularly in the states of Rajasthan and Gujarat.

Fried's argument, therefore, represents neither a definite type of society nor a definite stage of evolution. He defines tribe as a kind of secondary phenomenon:

> I shall go on to say that most tribes seem to be secondary phenomenon in a very specific sense: they may well be product of processes stimulated by the appearance of relatively highly organized societies amidst other societies which are organized much more simply (Beteille, 1999: 65).

Andre Beteille has rather emphatically expressed his preference for the historical as against the evolutionary approach in the definition and identification of tribes. In his opinion, in a country like India where tribe and civilization coexist, being a tribe has been more a matter of remaining outside of state and civilization whether by choice or necessity, than of attaining a definite stage in the evolutionary advance from the simple to the complex (Xaxa, 2003: 376).

The study of groups that later came to be described as tribes began with the establishment of the Asiatic Society of Bengal in 1874. Vidyarthy (1982) divides the phase of research on tribes in India into three phases (quoted in Xaxa, 2003: 373). The period from 1874 to 1919 is considered by him as the formative period. He identifies the later two phases as the constructive (1920-49) and analytical phases (1950 onwards). In the post-Independence period there was a flood of literature which came from three main sources - the Anthropological Survey of India, university departments of Sociology and Anthropology and tribal research institutes that were set up in states having a substantial tribal population.

Virginious Xaxa says that according to the opinion of some anthropologists the categories of caste and tribe are just 'colonial constructions' in the sense that the character of these groups was solidified by the British through the processes of classification and enumeration (Ibid. 2003: 374). There were different social groups corresponding to those identified as tribes in pre-colonial period too. But such groups which had distinct local and regional nomenclatures (for example Santhals or Nagas) were not categorized together under the general category of 'tribe'. In this sense, the category of tribe is part of the modern consciousness brought into being by the colonial state and confirmed by its successor after independence.

When the British began writing on Indian society, the term tribe was used in more than one sense. One sense in which it was used was to refer to a group of people claiming descent from a common ancestor. The other was to refer to people or communities living in primitive or 'barbaric' conditions.

In the 18th century, the terms 'tribe' and 'caste' were used synonymously. In the census undertaken in 1891, the term used was not tribe but 'forest tribe' (ibid. 2003: 375). Attempts have been made in the later censuses to arrive at a distinction between tribe and caste but the criteria for that either proved to be ambiguous or inadequate.

Though the terms 'tribe' and 'caste' were used synonymously, the British treated the segment of population later known as tribes as different from the rest of the Indian population. Thus for administrative purposes, groups came to be treated as tribe, for which separate laws were framed for their governance. Jaganath Pathy in his essay 'Tribal and Indigenous Peoples of the World', rightly says "...who is a tribal depends much on the administrative definitions adopted by different government" (1992:13).

But in the case of anthropological study, a tribe is viewed differently as a whole society with certain features. From some anthropologists' viewpoint, such as Honigmann (1966) a tribal society seems to be more or less homogenous with common government, language, culture, customs etc (quoted in Xaxa, 2003: 375). For some others, like Sahlins (1961), tribes are distinguished by the absence of state like features in their social organization which is marked by segmental features (ibid. 2003). Godelier (1977) views tribes as a stage in the evolution of society while for Fried (1975) tribe is only a kind of secondary phenomenon acquiring its form and identity from some external source (ibid. 2003).

In the Indian context, the classification between tribal and non-tribal is administrative rather than anthropological. Hence tribes are treated as those that are enlisted in the Indian constitution in the list of Scheduled Tribes. The present list shows 461 tribal communities, of which 174 have been identified as subgroups. The 1991 census estimated their population at 67,758,000 constituting 8.01 percent of the total population (ibid. 2003: 379).

Taking into account all the preceding views of

anthropologists and sociologists regarding the word 'tribe', we can finally arrive at a broad definition of it. We may also look at B. Saraswati's definition of tribe as referring to a

> ...constellation of somewhat reformed aspects of primitive life, generally constituting a homogenous unit, speaking a common language, claiming a common ancestry, living in a particular geographical area, lacking in scientific knowledge and modern technology, and having a social structure based on kinship (1991: 13).

The point to note here in the definition is that tribe in relation to modern society is "a separate humanity" (ibid. 1991). However, I have reservation about accepting the definition in its entirety, particularly in relation to the idea of lack "in scientific knowledge and modern technology". The parameters for judging the levels of progress of scientific knowledge of different societies are not clear. After all, there are no universally acceptable parameters to measure scientific knowledge.

In an interaction with the students and faculty members of Arunachal Institute of Tribal Studies at Rajiv Gandhi University on 20th April 2007, Andre Beteille said that a tribe is a *more or less* self-contained group with its own distinctive customs, often its own distinctive language and which requires its members to marry within itself. I feel that Beteille's definition of tribe would probably be the most acceptable of the lot because of its flexibility and comprehensiveness.

Section B

Arunachal Pradesh, the land of the Rising Sun, is inhabited by twenty six major_tribes and as many as one hundred and ten sub-tribes and about one hundred dialects (roughly) (Dawar, 2003:1). It is bounded by Bhutan on the West and Tibet and China on the North. On its South-East lies Burma whereas Assam stands on the Southern boundary. As per Arunachal Pradesh Human Development Report 2005, the total population of Arunachal (including non-tribals) is 1,097,968 (2006: 8). The ratio of male and female citizens of the state is 1000:893 (ibid. 2006: 93). They are scattered over an area of 83743 sq. kilometers throughout hilly terrains with rugged and typical mountainous topography. According to the 2001 census, the total number of people belonging to the Scheduled Tribe is 705,158, constituting 64.22% of the total population. The non-tribal population of the state is 392,810, constituting 35.78% of the total population (ibid. 2006: 8). They are spread across the three thousand six hundred and forty nine villages of the state (Pandey, Duarah and Sarkar, 1999: iii). Arunachal Pradesh is the largest state in the Northeastern region of India in terms of area. Till 1971 the state of Arunachal was known as the North East Frontier Agency (NEFA) and on the 21st of January, 1972 the NEFA was granted the status of a Union Territory and was christened Arunachal Pradesh. Subsequently, on the 20th February, 1987 this union territory was declared a state (Dhar and Coomar, 2004: 2). The state is a canvas of different tribal communities with their habitats amid virgin hills, dense forests, attractive greeneries and roaring rivers with a congregation of diversified socio-cultural beliefs and practices.

North-eastern India is one of those regions in the country that has exhibited accelerated population growth - both natural and artificial increase after independence (Gopalkrishnan, 1991: 24). Arunachal Pradesh is fortunate as it has indicated possible potentials of rich resources; large areas of land that have been underexploited and underutilized. The population in the region is unevenly distributed and it is very sparse.The state has the lowest density of 13 persons per square kilometer.

Agriculture is the mainstay of the livelihood of the people of Arunachal. The economy of Arunachal is largely based on agriculture. Every community has its own way of meeting its basic needs for the existence of its members failing which they are threatened with extinction. The first and foremost characteristic of the tribal economy is the close relationship between their economic life and the natural environment or habitat which is, in general, the forest. There has been a great change in tribal economy all over India, though not immediate or sudden. This is also true in the case of Arunachal. The line of change can be summed up in Vidyarthy & Rai's (1976:142) words:

(i) From forest hunting economy to forest hunting and agriculture,
(ii) From hill cultivation to settled cultivation,
(iii) Simple agricultural economy to multi-crop agricultural economy, labour and white collar job and commercial economy,
(iv) From artisan type to artisan plus marketing.

The people of Arunachal are basically agriculturists and their entire socio-economic life is dependent on cultivation. The system of cultivation is mainly slashing and burning followed by ploughing with the exception of the Apatanis who depend on wet rice cultivation. Rice is the staple food though millet and maize are also taken by some. They grow pineapples, oranges, banana etc. which are now marketed outside Arunachal. All the people have a tradition of artistic craftsmanship which manifests through the various products they produce. A wide variety of crafts which include weaving, painting, pottery, smithy, bamboo and cane work, wood carving and basketry are found among the people of Arunachal. Women are the main producers both in the field and at home (except in basketry) and the maintainers of the family. Of course, there are some who earn livelihood as employees of the administration, para-government bodies, contractors and traders. They live outside the village society and are fast forming an urban group. They represent a class which is without any stable mooring either in the village or in the town. The tribal societies were called casteless but had classes such as chiefs, commoners, slaves and freemen (Dutta, 2004: 13).

Section C

The North East India comprising of the states of Assam, Arunachal Pradesh, Meghalaya, Manipur, Nagland, Mizoram, and Tripura is the homeland of a large number of people of different ethnic groups with both tribal and non-tribal identity. Arunachal Pradesh, the largest state in the North East in terms of area, was originally inhabited by tribal people alone. Presently the total population is above ten lakhs and more than one third of this number consists of non-tribal people from different parts of the country. The tribes are scattered over the rugged mountainous terrain of the state and have difference in language, dress and customs. They have their own style of living, socio-cultural heritage, religious faith and belief, and ways of life. The various major tribes of the state of Arunachal are Monpa, Sherdukpen, Nyishi, Adi, Aka, Khowa, Miji, Apatani, Hill Miri, Tagin, Mikir, Memba, Khamba, Mishing, Mishmi, Khamti, Singpho, Zakhring, Nocte, Wancho, Tangsa and Lisu (ibid. 2004: 5). Of course the very idea of differentiating tribes as "major" or "minor" is questionable as obviously such categorizations are deeply political. There is no clear-cut criterion that is followed in differentiating between "major" and "minor" tribes.

Broadly the people of Arunachal Pradesh may be divided into three cultural groups on the basis of their socio-religious affinities (Elwin, 1958: xii-xiii). The first group of tribes professes Buddhism. They are the Monpas and Sherdukpens of Tawang and West Kameng district, Membas and Khambas who live in the high mountains along the Northern borders and Khamptis and Singphos inhabiting the eastern part of the state. They are said to have migrated from Thailand and Burma long ago and are still using scripts derived from their original homeland.

The second group of people is Adis, Akas, Apatanis, Nyishis, Mishmis, Mijis, Khowa, Sulung, Hill Miri and

Tagin etc. These tribes traditionally had their own belief-system. They worship the sun and the moon, namely Donyi and Polo and Abo-Tani is the original ancestor for most of these tribes. Apart from this, they are worshippers of spirit and soul. Their religious rituals and festivals largely coincide with phases of agricultural cycles. They invoke nature deities and make animal sacrifices in order to guard against diseases.

The third group comprises Noctes and Wanchos, agricultural communities, adjoining Nagaland in the Tirap district and the Tangsas. They worship nature, spirit and soul. The two tribes, Nocte and Wancho are included in the same group because of their association with the cult of head hunting. They are known for their strictly structured village society in which hereditary village chief still plays a major role.

However, Parul Dutta and Syed Ishteaque Ahmad, the editors of volume XIV *Arunachal Pradesh* of the People of India Project have divided the people of this state into five cultural zones, namely- i) the Mon cultural zone, ii) the Nishi cultural zone, iii) the Adi cultural zone, iv) the Mishimi cultural zone, and v) the Nocte-Tangsa cultural zone (Dutta & Ahmad, 1995: 20-21).

It is important to note that many other tribes of North East India, the tribal people of Arunachal Pradesh are passing through a crucial phase of transition. Prof. A.C. Bhagawati says, "The significant changes in the tribal pattern of life and culture resulted from conversion to Christianity and educational and other welfare measures introduced by the missionaries" (1992: 495). It is possible to divide the tribal people of Arunachal Pradesh into two cultural groups as Budhhists and non-Buddhists on the basis of their religion. But, Christianity has become increasingly popular in Arunachal Pradesh in recent years. The social situation of Arunachal Pradesh has undergone transformation. This has led to the emergence of many other issues like that of identity formation, inter-ethnic relations, as well as institutionalization of their religious belief-systems. The process of conversion can be explained in terms of "the technology the missionaries involved for the material well-being of the people. The development of written forms of various languages, printing of books (beginning with the Bible), medical work and opening of schools were but some of the important introductions into the tribal arena" (ibid. 1992: 494).

Different tribal groups have tried to establish their own identity on the basis of common language, culture, territory and self-consciousness with a set of common goals and objectives. And they often develop a sense of pride in their own identity. In the North Eastern states of India, especially in Arunachal Pradesh, the process of identity formation has some interesting features. Some major tribes try to include some of the smaller tribes in their groups to increase their numerical strength. On the other hand, some other tribes, or so called sub-tribes try to exclude themselves from the larger generic umbrella in search of their tribe-specific identity. For instance, earlier Galos were part and parcel of the Adi constellation – the second largest tribe, but within a decade there was a movement initiated by the some of the Galo intellectuals opting for a separate or distinct Galo identity which has strained the interethnic relations to such an extent that it gave birth to serious conflicts which were diffused with the intervention of eminent members from both the sides (Chaudhuri, 2007).

Thus, through negotiating intertribal relations, the tribal people are trying to redefine their identity. And this attempt to redefine identity and its assertions may be viewed as a part of the overall process of social change. The significant change in the socio-religious aspect of Arunachal Pradesh is not limited to the conversion into Christianity but to counter its phenomenal increase in this area, there is institutionalization of indigenous religious forms. Donyi-Poloism was initiated first within the Adis to promote indigenous faith which is a new social formation. Earlier most of the tribes were broadly bracketed as the believers of animism. Later on, to preserve and promote indigenous faith various societies such as Indigenous Faith and Cultural Society, Nader Namlo, and Nader Nele have been established. It is important to note here that Late Lummer Dai "had founded Donyi-Polo Gangging at his residential complex, Naharlagun where Donyi-Polo devotees used to gather every Saturday to offer their prayers, and also a grand function is organized on December 31st every year to mark the Donyi-Polo day" (Biodata circulated by Nanni Dai).

But in spite of the awareness of the traditional ethnic identity, it is strongly felt that quite a good number of tribal people are particularly concerned with the regional identity of Arunachal Pradesh as a whole. Prof. A.C. Bhagawati says:

> ... There are other issues, arising from time to time, when the identity framework is in terms of being an Arunachalee. A regional, state level identity is, therefore, at this moment something that can be called 'episodic identity'. It surfaces, no doubt quite strongly, as and when a particular episode arises, e.g., the All Arunachal Pradesh Students' Union led demands for demarcation of boundary with Assam, renovation of trading and contract licences of non-Arunachalese or expulsion of foreign nationals from the state (2004: 186).

In the proposed work of study, I try to look into the issues related to various aspects of Arunachal Pradesh as reflected in the novels of Lummer Dai and Yeshe Dorjee Thongchi. The issue of identity formation, both the individual tribe-specific identity and the regional identity of the Arunachalee comes under the purview of my discussion.

It is necessary to discuss the different tribal societies of Arunachal Pradesh before I discuss the literary writings of Lummer Dai and Yeshe Dorjee Thongchi. It is vitally important to have a fair idea of the societies in which the novels of Dai and Thongchi are set in order to be able to analyze them from a New Historicist perspective. Here is an attempt to briefly look at the socio-cultural order of those tribes on which some written materials are available.

1. The Adi

"The word 'Adi'", Verrier Elwin says, "covers a large number of tribal groups, united by a language that in spite of dialectical variations is fundamentally everywhere the same, and by a similar culture and temperament" (1959: 18). Tai Nyori divides the Adis into two broad groups: (a) Padam Minyong group consisting of ten sub-tribes - Padam, Minyong, Pangi, Shimong, Ashing, Pasi, Karko, Milang, Tangam, Bori and (b) The Galo group comprising of four sub-tribes - Galo, PaLibo, Ramo and Bokar (Nyori, 1993: 4). Each of these groups has its own style of living, social and cultural behaviour, economic pattern, yet there are certain features common to all the tribes.

The Adis were earlier known by the word 'Abor', "a rather derogatoy word meaning 'unruly' or 'disobedient' which correctly described the idea which the plains people had of them in the past" (Elwin, 1959: 17). The origin of the word 'Abor' has been interpreted differently. One interpretation of it says that it is originally an Assamese word meaning 'savage', 'independent' or 'hostile'. A second interpretation, put forward by D.P. Chaudhury says that it may have "some connection with Abo, the first man, according to Adi mythology, to whom they trace their origin" (quoted in Dawar, 2003: 7). Today, the Adis prefer to be called as 'Adi' which means 'hill' and thus in its extended sense, the hill man. Elwin says, "They themselves have suggested and the Administration has accepted, that they should be called Adis or hillmen" (1959: 18).

The Adi villages are situated on the hill spurs and houses are generally built of bamboo or wood and raised about 1.20 to 3.60 metres above the ground (Dutta Choudhury, 1994:112). The main part of the house is a big hall which serves as a bed-room, living-room and also dining room. The fireplace is in the middle portion of the house.

The Adis are basically agriculturalist and their entire economic life centers around land. *Jhum* which is known as *tumperike* (Borgohain, 2004: 94), is a common method practiced by them. Terrace rice cultivation (TRC) and wet rice cultivation (WRC) are also practiced by them. Before 1950 there was no permanent cultivation and *jhum* was the only system of cultivation. Like agriculture, fishing in the river is also an important part of the economy and passtime of the Adis and the areas for fishing are well demarcated. Every village has its own hunting-grounds and fishing areas (ibid. 2004). Like the possession of cultivable land, the fishing sites are also purchased or inherited from the father. The ownership of a small stream goes to the family who first located the stream or the ownership is enjoyed by a family through whose land it flows.

The non-cultivable land normally found at the top of a hill belongs to a particular village within its boundary. It is normally kept reserved for hunting games. Thus each village has a common hunting ground. Besides this common hunting ground, a tribe like Minyong has their individual hunting grounds in their own *jhum* lands. There are some trees like 'Paku', 'Pake' and 'Balam', the seeds of which are eaten by the wild animals like deer (Hodum), wild bear (Hore), and other villagers are not supposed to kill the animals grazing under these particular trees (ibid. 2004: 98).

The Adis celebrate a number of festivals with pomp and gaiety. Each festival has a distinct purpose for which it is celebrated. Of the popular and traditional festivals of the Adis, Mopin and Solung are the most important. These are agricultural festivals celebrated respectively in April and in September for abundant crops and prosperity of the people. The rituals are performed to propitiate the deities. Mopin is celebrated more particularly by the

Gallong group (Dutta Choudhury, 1994: 87) for five days. On the second day of celebration, a mithun is sacrificed and it follows a group dance to the tune of *poper-tingum,* a ballad narrating a myth. As the dance goes on, each of them present there smears the face and hands with rice powder (ibid. 1994: 88). Solung is one of the most colourful festivals of the Adis and is celebrated for seven days. During these auspicious days they prepare rice beer and plenty of meat and vegetables to entertain guests and family members. Their traditional dance called *ponung* which is very popular all over Arunachal, is performed on this day.

The Adis believe in the existence of supernatural beings and innumerable malevolent and benevolent gods and goddesses, spirits and deities. "They composed myths to seek an explanation of the mysteries of nature, which they endowed with a spiritual quality" (ibid. 1994: 76). The religion of the Adis comprehends a rich variety of ideas, beliefs and faiths which is profound, complex and subtle (ibid. 1994). A number of rites and ceremonies are performed by sacrificing animals and birds with the help of their priest called *Nyibu.* The Adi religion transcends into a far higher level of theology than the so-called "spiritism" or "animism" (ibid. 1994: 79). The Adis' religious thought and beliefs centre round Donyi (sun) and Polo (moon) and accordingly they have developed a religion called Donyi-Poloism.

The traditional political institution of the Adis is very strong. It is the village council called *kebang* which is responsible for maintenance of law and order in the village. It is of four types - *erang* or *opin kebang* for settling matters of the family and the clan, *delung kebang* for settling civil and criminal cases of the village, *bango kebang* for settling inter village disputes, and *bugum bokang kebang* for settling inter-village, inter-tribal and inter-distict disputes (Megu, 1999: 212-213). *Bugum bokang kebang* is the highest and supreme council responsible for dealing with all aspects of the society like peace, war, culture, religion, and development. "Another important feature of the Adi social organigation is the 'dormitory-club' for boys and men, called *moshup*...The dormitory is not only the meeting place of the youth of a village, but is also used for deliberation of the *kebangs*"(Chowdhury, 1979:148-149). Monogamy is the general rule of marriage, but instances of polygyny are not rare. The rule of clan exogamy is strictly followed by all, whatever the different forms of marriage may be (ibid. 1979: 100).

It would be worth mentioning here that Lummer Dai does not look at the divisions of Adis into different groups in his novels; he looks at the overall identity of the Adis.

2. The Sherdukpen

The Sherdukpens are one of the most progressive tribes inhabiting the southern part of the Bomdila range in the valley of Tengapani river of West Kameng district. They live mainly in three villages - Rupa, Shergaon and Jigaon, and are spread in approximately twenty hamlets or *lurek* (*lu* means hill and *rek* means cultivation). Some of the important hamlets are *Thungri, Mukuthing, Mingmachur, Jamyan, Brakpublang, Gacham, Rangthong, Diksi, Diringthing, Dharmail, Foot-Hill, Killen, Birpur, Soumkchi, Lingcham, Chillingpa, Lingbaktang,* and *Machakching* (Koley, 1999: 13). However, at present some of them have settled in Kamengbari area of Bhalukpung circle.

The Sherdukpen tradition claims that the tribe originally came from Tibet (Sharma, 1960: 5). According to the history of their migration, Be Srongtsen Gomoo, a king of Lhasa married a daughter of Ahom king. That queen gave birth to two sons. And the younger of them was Japtang Bura. It is believed that Japtang Bura was the first king of the Sherdukpens (Koley, 1999: 14). According to the tradition, the decendents of Bura now form into a class by themselves, known as Thongs, while decendents of his retinue are called Chhaos enjoying a slight inferior social status (Chowdhury, 1979: 48). As such the society is primarily divided into two classes- *Thongs* and *Chhaos*. Another group is called *Yanlo*. R.R. P.Sharma talks about the clan and sub-division of clans and says:

> Japtang Bura himself had three sons, named Thongdok, Thonga and Khrime, and these became the ancestors of three exogamous clans. Afterwards, there was further sub-division of three clans, e.g., Karmu came to be considered as a sub-clan of Thongdok, Musabee and Wangja of Khrime, and Thongchi and Lama of Thonga (1960: 49).

The Sherdukpen society is patriarchal. Monogamy is a feature of society. A Sherdukpen cannot have two wives at the same time under any circumstances. Class endogamy and clan exogamy are the essential features of their marital system. Marriage with the son of mother's brother or, conversely, with the daughter of father's sister is preferred (ibid. 1960: 54-55). Marriage by capture is also seen in the Sherdukpen society, particularly when the bride is unwilling, but her parents or the boy's parents are quite agreeable (ibid. 1960). Widow marriage is permissible.

They are Buddhists by religion having cultural affinities with the Monpas. The religion of the Sherdukpens is a "curious blend of Buddhism and local beliefs" (Chowdhury, 1979: 54). Lossar is the main festival of the people. In addition, two major festivals, *Chekar* (May-June) and *Kisscaba* (November) are celebrated by them with pomp and gaiety. The Sherdukpens have great love and interest for dance and pantomime. The most popular are the Deer dance, Ajilamu dance and Eagle dance. These dances in which they use masks are symbolic, depicting old legends or mythology (ibid. 1979: 55).

The grouping of houses in a village does not follow a definite pattern. The houses do not face the same direction. A typical house has two compartments with a portico in front. The room next to the portico is used commonly as kitchen, dining room and living room. The houses are generally double-storied. The lower apartment is used by the household while the important household belongings are stored in the upper compartment. Public gatherings are usually held in front of the house of the village headman who invariably occupies an elevated position in relation to others.

A Sherdukpen man wears a piece of cloth, called *sapo,* diagonally covering the upper part of the body. Two ends of the cloth are then secured on the shoulders (ibid. 1979: 49). A full-sleeved jacket, reaching below the hip, is worn over the *sapo.* It is open in front. Those who can afford put on short coats over the jacket or a long one, and loose trousers in winter. The Sherdukpen man cuts his hair short and puts on a skull-cap made from yak's hair. This cap is called *gordum* (ibid. 1979).

Alike other tribal societies there are village councils in the Sherdukpen community which bears traditional importance in the village. Sherdukpen village councilors are elected democratically by villagers in village forum. Village council is divided into ranks - the *gaonburha* followed by the active member, and, finally, the messenger. Members are selected considering their individual character and ability.

The Sherdukpens have traditionally collected taxes from the Kachari villages which they regarded as their own territory. Trade and barter play an important role in the economy of the Sherdukpens. On their annual migration to Doimara, they carry with them cattle, poultry, chillies, dried radish, the *jabrang* spice, daos and woven bags and sell them in the plains. In exchange they get *endi* silk, mill-made cloth, salt, rice, beads, bangles and cooking utensils made of metal (Sharma, 1960: 8).

3. The Nyishi

The Nyishis inhabit the vast stretch of hilly terrain in the Lower Subansisri, Papum Pare, and East Kameng districts. J. N. Chowdhury says, "The Bangnis occupying the eastern part of the Kameng district are really a western branch of Nishi tribe (formerly called Dafla)..., both the terms Nishi and Bangi signify 'man'..." (1983:26).

Chowdhury says here that he has not, however, come across any explanation as to why the western branch of the tribe is specially designated as Bangni. And he says that apart from their geographical situation the Bangnis differ little from the Nyishi tribe to their east and they are in reality the same people (ibid. 1983: 27). Bibhash Dhar says, "The Bangni people, like the Nishis, were known as Dafla in the past ... The Bangnis are very close to the Nishis ... They are recognized as a Scheduled Tribe under the fold of the Nishis" (1995: 147). Parul Dutta says, "The Bangnis and the Nishis are the same group of people found in the East Kameng and Lower Subansiri districts respectively" (2004: 6). They have their own dialect which belongs to the Tibeto-Burman language group. The Nyishis believe that they descended from Abotani - a mythical ancestor. He begot three sons, who, in their turn, are believed to have been the forefathers of three clans namely Dapum, Dodum, and Dol: "All the groups of Daflas/ Bangnis/Nissis/Nissang trace their past to one ancestoral group which had three groups of Dapung, Dodum and Dol" (Gopalkrishnan, 1994: 55).

The Nyishis were earlier referred to as 'Dafla', a term no longer in use for the derogatory connotations it has acquired over time, "The so-called Daflas have completely discarded the name as both foreign and derogatory, and insist on being called 'Nishi', a term said to be native to the tribe" (Chowdhury,1979: 81). Haimendorf says in this context:

> They (the Nyishis) dislike the name Dafla because in Assamese it has the derogatory connotation "wild man" or "barbarians" (1980: viii).

The Nyshis do not live in a village community but as number of families bound by kinships, live together in unusually long houses. A single long house might shelter twenty families at times. Verrier Elwin says;

> They have very long houses occupied by a number of related families...though there may be several

> such houses in a single settlement; they are not usually united into a village community (Elwin, 1959:13)

Thus, the real social unit for them is house. Each family has a hearth and all the affairs of the family are centered round this hearth. The eldest male member of the house is respected by all and he also exercises his command over all members. Payment of bride price is prevalent in the Nyishi society. The society never recognizes any marriage if mithun is not offered by the bridegroom. The minimum bride price is one mithun and it may go up to twenty mithuns. Polygyny is very common (Chowdhury, 1979: 86). Each wife has a separate hearth and also a separate plot of land which she cultivates.

The Nyishis are used to shifting cultivation. They are not self-sufficient in food grain production. The Nyishi women folk is expert in weaving and the Nyshi men folk is skilled in cane carving. They rear cow, mithun, pig, goat and poultry birds for the purpose of meat.

The most distinctive part of the dress of a Nyishi is his helmet, surmounted by a red-dyed beak of hornbill. It may also be decorated with hornbill feathers. Just below the helmet, pushed a little backward, the bun of plaited hair called *podum* protrudes over the forehead (ibid. 1979: 83). A piece of yellow strap is tied round the *podum* and a brass skewer is stuck through the podum horizontally. The lower garment consists of a loin cloth reaching halfway to the thighs. They wear multicolored beads round the neck. But the whole outfit is incomplete without the sword hung from the neck by a string.

The women gather their hair in a bun at the back of the head with a simple bamboo comb. The traditional garment is a skirt of woven fibre with green border or stripes. The waist is decked with cane belt decorated with *hoofi* (ibid. 1979: 84) metal disc. A *tage* (ibid. 1979) or chain with a number of flat square metal pieces and blue beads is worn round the neck. The ankles are encased in tight fitting cane garters.

The major festival of the Nyshis is Nyokum. It is celebrated on 26th February every year to propitiate the gods and spirit for a bumper harvest, welfare and prosperity. It is an agricultural festival observed just before the beginning of Jhum cultivation. The priest performs rituals by offering animals and fowls in front of the altar especially erected in the festival ground. A festival feast is also organized. Folk songs and dances are perfomed.

The Nyishis bury the dead bodies and the purification rites are peformed by sacrificing animals and birds. They perform a number of rituals in a year with the help of local priest called *Nyibo*. They have a number of dances which they perform on various occasions. The most popular dance is Rikhampak which is performed during Nyokum. Other dances are Northon, Buiya Sadan etc.The Nyshis practice Donyi Poloism. But at the same time conversion to Christianity is also seen.

All the disputes of a Nyishi village are settled by a council. The council is constituted with the gaon bura and he is appointed by the government. There is no specific number of the member of the village council (*Bang-Nyele*). *Nimai Yallong* is held for settlement of marriage dispute, *Dacho Yallong* is held for the settlement of theft cases. *Marana Yallong* and *Mingrana Yallong* are for the quarrel and murder cases; similarly a, *Kady Yallong* and *Yoshi Yallong* are held for the settlement of the land disputes and sexual offences respectively (Srutikar, 1999: 78).The role of go-between or *gindung* (*But ayendo*) is remarkable. The *yallong* consists of three members: *nimai yallong, decho yallong,* and *yello tolla*.

4. The Monpa

The extreme western part of Arunachal Pradesh is the homeland of the Monpas, a Buddhist tribe of Mahayana sect. The Monpas live in the north-west part of the West Kameng district, that is, within the radius of five miles of Tawang, the valley of Dirang and the south-west of Kalaktang region. The Monpas are divided into three sections- Tawang Monpa, Dirang Monpa and Kalaktang Monpa (Bagchi, 1990: 37). They are supposed to have emigrated from Tibet through Bhutan. Dr Verrier Elwin describes the Monpas as "quiet, gentle, friendly, courteous, industrious, good to animals, good to children..." (Elwin, 1959: 10).

R.K. Deuri writes about the Monpas:

> The Monpas with 27500 souls according to 1971 census are a simple, gentle and courteous people. They profess Mahayana sect of Buddhism with perhaps an admixture of their original animistic beliefs. The cultural and religious life of the Monpas centre around the Tawang monastery (1983: 15).

One of the religious establishments among the Monpas is the *gompa which* is a temple to house the image of the Buddha and also religious books. The main concentration of the Monpas is around Tawang where stands the biggest monastery in India at an altitude of about 3,048 metres from the sea level, capable of housing 500 lamas (Chowdhury, 1979: 40). Scattered about the country side and in the village, are stone shrines known as *mone* and *chorten.* A *mone* is of two types, it may be either "a wooden cylindrical barrel fitted with a wooden shaft passing longitudinally through its middle and forming its axis (or)...in the shape of a stone wall of varying lengths" (Sarkar, 1980: 19). The setting up of a *mone* is considered as an act of piety and is believed to bring religious merit. Tablets bearing the sacred inscription, *Om mone padme hum* (Hail to the Jewel in the Lotus) are ranged on small shelves cut into the wall (ibid. 1980). *Chorten* is a *stupa*-shaped structure where prayers are held occasionally. The lama leads the prayer by sitting on the ground in front of the *stupa* (Sharma, 1960: 77).

The people live in houses constructed of stones with wooden planks, usually double storeyed. The garments of the menfolk are half-sleeved jackets called *chuba,* over which full-sleeved maroon coloured coat, *aliphudum* is also put on. The waist is tied with a cloth called *kichin*. Their female folk put on *shinka, tenga-khima,* along with *kichin*.

There are two ways of marriage in Monpa society. One way is the traditional system of negotiated marriage and the other one is love marriage. R. N. Bagchi says, "Marriage by abduction and capture is not in vogue in the (Monpa) society...and marriage with one's own cross cousin is permissible" (Bagchi, 1990: 49). Payment of bride price is a social custom. One woman may act as wife of two brothers (ibid. 1990). Divorce is traditionally forbidden in Monpa society. They are particularly serious in maintaining rituals related to the dead - for instance, the custom of slicing the dead bodies into a hundred and eight pieces (Chowdhury, 1979: 45).

Economically, this tribe is one of the more prosperous tribes of Arunachal. They are skilled weavers and agriculturists. They rear animals for economic benefit and use wool, milk and milk-made products. They utilize the agricultural land intensively and use animal driven plough. They have also used the hill-slopes, terracing them for cultivation. They use manure of jungle leaves, cow dung, pig, chicken dung etc.

They are very much influenced by the Tibetans, even in their way of life and food habits. For example, the tea with salt and butter and the *momos* prepared by the Monpas are well liked by the people who visit them.

The Monpas make beautiful masks. They also periodically stage pantomimes and mask dances. Making of beautiful carpets, painted wooden vessels and silver articles are, however, the specialty of the Monpas. They weave lovely colourful carpets with dragons, geometric and floral designs. The Monpas also make paper out of trees. This hand made paper is used for writing religious prayers and hymns.

The Monpas are very much fond of dance, music and arts. Their door and window frames are ample testimony of their art and they add to the beauty of their beautiful stone built houses. Their main dances are Yak, Lion and

Peacock, *Sonyi Yalo, Ajhi Lhamu* pantomime, *Broh* etc. *Ajhi Lhamu* pantomime is a popular dance form among the Monpas. Monpa and *Sonyi Yalo* are performed to welcome and honour guests and dignitaries during special occasions. The Lion and Peacock dances symbolize that prosperity comes when there is complete harmony and cooperation between all living beings on this earth. Yak dance is a popular dance in which masked dancers represent the members of a family who supposedly discovered the Yak with the help of a magical bird. Yak, being a multipurpose animal, is looked upon as the provider of permanent source of wealth and prosperity.

Lossar is the major festival of the Monpas in which the advent of the Monpa New Year is celebrated. It lasts for five days. Religious flags are hoisted at the top of the houses. Butter lamps are lit and people visit their friends and relatives during this festival.

The Monpas have a well-developed village self-governing institution for the last few centuries. They have a village council chosen by all adult members of the village. Among the elected members, the *tsorgan* (chief) administers the affairs of each group of the village. This way, the Monpas have a well developed body of customary laws and the council administers the common property of the village and also decides the criminal disputes. The literary meaning of *Tso Tsorgan* is 'the chief' who shoulders the responsibility of a group of villagers (Norbu, 1999: 1). While selecting a *Tso Tsorgan* the criteria like experience, ability, social status, knowledge of the traditional conventions, customary laws, oratory skills, integrity etc. are taken into account. The *Tso Tsorgan* is not paid any salary. However, he is given some remuneration in the form of free labour (*Blanpa*) for a specific task by the villagers (ibid. 1999).

5. The Apatani

The Apatanis are one of the major tribes of Lower

Subansiri District of Arunachal Pradesh. There are seven original villages (*lamba*) situated on the edge of the valley. They are Hari, Hong, Reru, Tajan, Jija, Michi, and Baming (Yapa, 1999:111).

The Apatani village is not just a conglomeration of houses, but reveals a definite pattern of arrangement. Each village is divided into separate quarters occupied by a specific clan. Within the village, social life and activities move around two focal points namely *nago* (a small ritual house) and *lapong* (the public platform). The Apatani family comprises of parents and unmarried children. The houses are built on raised platforms.

The older men-folk tie their hair in top-knots and tattoo the faces. Wearing of circular nose plugs and tattooing of faces is the most characteristic aspect of ornamentation of older Apatani women. However, new generations of Apatani men and women have stopped the practice of tying hair knot, nose plugs and face tattooing. The Apatani women wear coloured beads around their neck.

The Apatanis are good cultivators and practice both wet and terrace cultivation. Paddy cum fish cultivation is very popular among them. They utilize every possible bit of land for laying a rice terrace. They make extensive use of irrigation channels to keep their lands under different states of submersion. They have developed extensive indigenous methods of irrigation to water their fields from the natural sources. Being industrious, compact and highly organized tribesmen, they are mostly cultivators and they always produce surplus food grains. Unlike other tribes of Arunachal their economy is stable.

In marriage (*mida)* monogamy is common and they strictly adhere to clan exogamy and tribe endogamy. There are two classes viz. *guth* (the upper) and *guchi* (the lower) in their society and interclass marriage is strictly prohibited. In his book *A Himalayan Tribe: From Cattle to Cash* (1980), Haimendorf says:

> ...there is the horizontal division into two endogamous classes known as *guth* and *guchi*, two

> terms which may be translated as patricians and plebeians or commoners ... The division between *guth* and *guchi* is considered unalterable and so far neither wealth nor education and political success have enabled a *guchi* to rise to the status of *guth* except in rare cases of persons of mixed parentage (1980: 87).

A number of rituals and ceremonies are performed to appease the different gods and spirits. They consider Donyi-Polo (sun and moon) as their deity and thus practice Donyi-Poloism. The three major festivals of the Apatanis are Dree, Myoku, Murung etc. Dree is the main agricultural festival of the Apatanis celebrated for three days starting from 5th July after completion of plantation of paddy seeking bumper cropping, welfare and prosperity of the people and animal. The rituals are performed with the help of priests. Folk songs and dances are performed by both the sexes. Food and drinks are lavishly served. A community feast is offered to the guests and visitors.

The Apatani society is governed by some set of unwritten customary laws which are handled by the council of elders called *buliang* (Chowdhury, 1979: 120). There are various kinds of *buliang* i.e. *neha buliang, kidi buliang, kiimer buliang* and *muddo buliang*. Each clan of the society has their own *buliang*.

6. The Nocte

The term 'Nocte' is the combination of two words - 'noc' which means village, and 'te' which means people (Ahmed, 1995: 281). The Noctes are scattered in the south-western and the central parts of Tirap District. Namsang and Borduria are two important villages. The original religious belief of the Noctes centers around their god Jauban. But they have adopted Vaishnavism (Pandey, 1997:14). Most of them profess Vaishnavism and are disciples of the Bareghar Satra of Nazira, Assam (Dutta, 2004: 11). Noctes are famous as salt producers which is their chief item of trade and barter. They cultivate betel leaves on a commercial scale. They are agriculturists.

The society is not caste-ridden as such but they have the institution of chieftainship, the chief having great influence over the people. The society is actually divided into two broad classes but the division is not very rigid. The chief and his descendents form a separate class called Lowangjat and the commoners are regarded as one class known as Sanajat (Chowdhury, 1979:194). The two classes are again divided into many clans, making up each class, called Ku or Ru, but the terms may vary in different groups or villages. Tribe endogamy and clan exogamy are the general rules of marriage (ibid. 1979). In the past, head hunting was known and tattooing of special designs connected with it was common (ibid.1979:197).

They celebrate number of festivals. One of them is Loku. Renglo Loku is celebrated at the advent of agricultural session while Chalo Loku is celebrated on twenty fifth November every year. Pigs and buffalos are sacrificed with the help of the priests during Loku. It is celebrated for all round prosperity of the community. The famous Chalo and Loku dances are performed during these festivals. The most popular dance of the Noctes is Loku-Bawang.

The single family with parents and their married sons and unmarried daughters constitutes the basic units of the Nocte society. But after a child is born, a son cannot continue to live under the parental roof. The method of disposal of the dead by exposure on open platforms still has a vogue among the Noctes (ibid. 1979: 205).

The house style is of the usual *chang* pattern. They make use of wooden pillars as the framework for the house. *Toko* leaves are used for thatching. The original dress of the man consists of numerous cane belts round the waist, a strip of cloth worn in the style of *langhoti* with the rear end hanging and bamboo leaves worn round the legs and arms. The women wear a piece of cloth round the waist usually of white or black colour reaching down to the knee.

7. The Aka (Hrusso)

Aka is the name given to small tribal groups inhabiting the southern area of East and West Kameng Districts. The Aka area is bounded on the North by the Miji territory, on the South by the Sonitpur District of Assam, on the West by the land of the Sherdukpens, and on the East, by the Nyishi tribe. 'Aka' literally means 'painted'. It is obviously an Assamese word which might have been applied to a tribal group who called themselves Hrusso because of their custom of smearing their faces with black resin obtained from the *pinus excelsus* and charcoal (Chowdhury, 1979:55). The prevalent belief among the people is that they migrated to their present habitat from a place known as Ziga (Lonkapur), somewhere in Upper Assam (Ahmed, 1995: 135).

The social and economic life of the Aka is largely determined by the conditions of their habitat. They are mainly jhum cultivators. The abundance of jungle provides ample resources to practice their indigenous system. The abundance of bamboo provides them the material for house building. A number of pots and vessels of their daily use are made out of bamboo. The house is a work of bamboo and wood, bamboo sheets forming the floor, the walls, and the roofs and the wooden logs serving as the main pillars.

The Akas are very hard working. Baskets of different shapes and sizes are made by the menfolk. The most common form of Aka art is their drawings on wood in their house which are drawn with the liquid of *ling-hong* (pine resin), the essence of which gives black dye. Their staple food is maize.

Men wear a small garment hanging below the shoulder and covering the upper half of the body. It is known as *oposhiu-gu*. They also wear a long coat known as *pol*. Women wear a dark red garment over the body hanging from shoulders up to knees also known as *pol*. Both men and women cover their legs with a piece of cloth called *gudu*. The common silver ornaments of Aka women are *gichli* (the earrings), *jeggul* (wristlet), and *aescheri* (coloured bead necklace). The women tattoo their faces in a pattern of straight line.

Sadinaktso, Neichu Dou, and *Dodou Dou* are some of the Aka dances. In *Sadinaktso* one member each from the groups of boys and girls simultaneously give their performance. No song is sung. *Neichu Dou* is a happy dance performed by the Aka youths. They sing and dance collectively glorifying the beauty of flowers and omnipotence of the creator. *Sarak* is one of their festivals.

The village council of the Aka is called *Melley* which is presided over by the *gaonburha* or the headman. There are two other functionaries, called the *borah* and the *gibba,* who are members of the council (Chowdhury, 1979:62).

Slavery at one time formed a prominent feature of the Aka society. The slaves (*khula*) constituted a class by themselves. The principle was once a slave forever a slave. The treatment of slaves was 'humane' and 'considerate' (ibid. 1979: 65-66).

8. The Bugun (Khowa)

The Khowas are a very small group of people living among the Akas but in separate villages. They call themselves Bugun. They belong to the Mongoloid stock and have their own language. The Khowas are at present spread over eight villages on the mountains on both sides of the Rupa river, in the West Kameng district of Arunachal Pradesh (Dhar, 1995: 185).

"The Khowas have had links with the neighbouring Sherdukpens and Akas since a long time. It is said that they used to provide labour to the Sherdukpens in exchange for clothes" (ibid. 1995: 188). They are good cultivators traditionally and labour for the Akas. Usual process of cultivation is jhuming. They are self-sufficient in food. They consider it a disgrace to be dependent on others for food grains. They rear mithun, goat, pig, fowl etc. Mithun is used as currency for paying bride price and its meat is eaten on all occasions for feast.

The family comprises of parents and unmarried children. The houses are double-storeyed, constructed with cane, bamboo and wooden posts. The society is patriarchal and patrilineal. In marriage monogamy is the basic rule but polygamy is also practiced. Widow remarriage is permitted. Tribe endogamy and clan exogamy is strictly followed.

Their religion maybe called animistic, consisting of sacrifices to various spirits (Chowdhury, 1979: 72). They believe both in malevolent and benevolent deities. They dispose of their dead body by burial. The Kishwt Sowai is the annual agricultural festival celebrated in the month of January. It is a ceremony about the benevolent deity Chamram who as they believe has a partiality for everything white including the animal to be sacrificed. No villager is allowed to go out and no outsider is allowed to come inside during the festival. Each village has a traditional village council.

9. The Wancho

The Wanchos inhabit the extreme eastern part of the state and in the south-western part of Tirap District, bordering Nagland. They are carefree, cheerful and hardworking people. Tattooing is a social custom among them. They believe in the existence of two powerful deities, Rang and Bourang. The women are good weavers but the art is restricted to the members of the chief's families only. They are expert in wood carving also. The dormitories are the centres of art and craft of the people. "The dormitories are decorated with buffalo and other animal heads killed on various occasions, and human heads possessed through head hunting expeditions" (Barua, 1990: 79).

The Wanchos celebrate a number of festivals with gaiety and fervor. One of them is Oriah. The other name of Oriah is Ojiayale. It is an agricultural festival celebrated on sixteenth February every year community basis for four to six days to propitiate the benevolent gods. A number of birds and animals are offered in the name of concerned gods with the help of priests. On the concluding day, men folk get together in the dormitory for community feast and drink. Most popular dance of the Wanchos is Ozele. They believe in Donyi-Poloism.

The Wancho villagers are grouped into several confederacies (*jana*) and is under the control of the chief. The *jana* is also divided into a number of units called *jang* and these are administered by a junior chief.

One of the characteristics of the Wancho society is the existence of class system. There are four classes - Wangham (chief), Wangshas, Wangsus, and Wangpans (Dutta, 2004: 35). The chief occupies the top position. The Wangshas are the offsprings of Wangham fathers and Wangpan mothers and Wangsus are the children of Wangsha fathers and Wangpan mothers. Wangpans are at the lowest rung of the social ladder (Chowdhury, 1979). The council of elders is called Wangchu-Wangcho or Ngajow council. The institution of chieftainship is hereditary. The social distinction of the classes is not visible in the normal course of life but easily recognizable during social functions, rituals and festive occasions. In marriage polygamy is permitted and cross cousin marriage is preferred. Well designed beads and necklaces are common among them. It has both social and ritual value. From the designs of tattooing one can easily identify the position of the men and women in the society. The women also blacken their teeth for decoration. The Wanchos are basically agriculturists and practice jhuming.

10. The Khamti

The Khamti is a Buddhist tribe of Himayana sect. They inhabit the lower region of the Lohit District. They call themselves Tai or Tai Khamti denoting that they belong to the Tai race. They derive their name from the place inhabited by them i.e. 'Kham' for cold and 'ti' for place.

Their language belongs to the Tai speech family and has their own script derived from Shan (Pandey, 1997:17).

The Khamti villages are of permanent nature having well demarcated boundary. Each village has a Buddhist monastery called Bapuchang. The houses are constructed with a specific pattern on raised floors with strong timber structure and thatched roofs. Polygamy is socially recognized and in such marriages the first wife always exercises authority over the other wives. Bride price is prevalent.

The women are expert weavers and most of the garments are woven by themselves. Rice is their staple food. They are settled cultivators and use bullock for plowing the field. They rear animals like elephants, buffaloes etc.

They have a well organized system of self-government. The institution is called Mokehop or Mok Chup Khai Kham which is headed by the village chief (Chowdhuri, 1999: 315). The members are drawn from the people of various status. The administration of justice is said to be regulated by their sacred text known as Thamasat. The Panchashila is an important aspect of the Buddhism followed by the Khamtis. In every religious prayer the Panchashila is pronounced. The Sanken is their main festival celebrated in the month of April every year. The birth and death anniversary of Gautam Buddha are also celebrated (Pandey, 1997: 18).

The Khamtis are good agriculturists. The whole community joins together in cultivating lands on a commercial basis. Animal husbandry is quite advance among them. They keep buffaloes and oxen both for tillage of land and for barter trade.

The Khamti society is divided into classes signifying distinct status in the social hierarchy. The chiefs naturally occupy the highest position, followed by the priests who wield very considerable influence over all ranks. Next comes the common freeman. Then comes the slaves who constitute the lowest rank. Manual work was done only by the slaves (Chowdhury, 1979: 182). The method of disposal of the dead is by burial.

11. The Lisu (Yobin)

The Lisus are a small group of people inhabiting the remote eastern most corner of the Tirap District. They speak a language called Lisunga. "The Lisus describe themselves as 'wandering people', which indicates that they were of nomadic habit" (Dutta, 1990: 14). They weave their own garments at their looms. Men wear a shirt reaching above the knees and a long overcoat and bags of langoor, jungle cat, cow and other skins are in fashion. A turban and a waist belt are also worn. Men carry a sheathed sword tucked in the belt and a crossbow with a bag of arrows hanging from the shoulder when they go for hunting. The female dress consists of a black jacket by the women and a loose skirt or petticoat (*jamubachi*). The man wears a *jabachi* (coat) (Maitra, 1995:190). Married women put on an additional piece of cloth over the frontal part of petticoat. They also wear a double row of white buttons round the head from which strings of beads hang in a fringe, terminating in a tiny metal belt. They also wear bone ornaments around the head. They wear necklaces, bracelets and earrings of white, red and blue beads.

The Lisus live in well demarcated villages, each comprising a number of houses. The houses are built on platform above the ground. The floor and walls are made up of split bamboo tied with cane. It has doors but no windows. Every house is divided into two parts - one is used as a guest room, and the other for dwelling purpose.

The Lisus are divided into a number of clans according to their profession. They are Ngawaja (representing fisherman), Jeja (representing beekeepers), Leme, Mise, Miko, Jali, Naphu, Tache, Hejapha, Mubipho etc. They are patrilineal. A Lisu family generally consists of the parents and the unmarried sons and daughters. The younger son stays with parents and the property, both moveable and immoveable, is generally inherited by the youngest sons after the death of his father. Divorce is rarely found. Burial is the traditional method of the disposal of the dead. The dead body, laid in a coffin, is buried in the burial ground on the outskirts of the village.

12. The Tangsa

The Tangsas, one of the major tribes, inhabit the Changlang District along the Indo-Myanmarese border. Tangsa is a common name of a group of people consisting of the Lungchag, Moklang, Yugli, Lungri, Mashong, Takhak, Ponthi, Kimsings, Ron-rangs, Longphi etc. Each group is subdivided into a number of exogamous clans. Racially, they are of Mongoloid stock. The literary meaning of the word 'Tangsa' is 'hill people', 'Tang'-hill, 'sa'-people (Pandey, 1997:13).

There is no class distinction among the Tangsas. They are monogamous, have patriarchal families, and live jointly. The existence of the Tangsa village council dates back since time immemorial. The council is constituted of the village headman and other elderly influential persons of the village. The council is informal in character.

The house type is the usual pile dwelling prevalent in this area with some variations in the layout. A house has three main parts. The front part is open except for side walls. This front room is called *ran* and used both as a common room and a guest room. It has a fireplace (*tap*). The mid portion, divided into a number of small compartments is called *jumong* or *himong* and the back part, where the paddy pounder is installed is known as *jumchang* or *himkha*. Guest of other sub-tribes should not go beyond the *ran*. The walls are decorated with skulls of buffaloes, pigs and other animals. The Tangsa man wears a *lungi*, a skirt and a turban. The original dress consisted of a loin cloth (*rai*), a sleeveless shirt (*santong*) open in front and a piece of cloth (*khupak*) for the head, all manufactured by them. The traditional dress of women consisted of a *khasa* or *khatsa*, a single piece of cloth wrapped round the body from the breast down the ankles, a blouse, and in some cases turbans. Tangsa men wear coloured beads in their ear lobes held by copper wires. They also put on cane leggings. Earlier they wore cane waist-bands which have gone out of fashion now. Women wear a transparent crystal bead in their ears known as jangphai. They are expert in cane and bamboo basketry work.

The Tangsas believe in spirits. There are different methods for the disposal of the dead among the different sub-tribes. While some of them bury the dead, some others cremate.

13. The Singpho

The Singphos live in both Changlang and Lohit Districts. They are Buddhist by religion and profess Mahayana sect. The society is divided into several classes, clans or small communities, each under a chief. The institution of chieftainship is also prevalent among them. But the position of the chief depends on his personal quality and ability. The chieftainship is hereditary and generally the youngest son succeeds (Pandey, 1997:15).

The Singpho family is patriarchal and patrilineal. In marriage clan exogamy is strictly followed. Payment of bride price is must. Monogamy is the rule but polygamy is also practiced. They prefer cross cousin marriage. Widow remarriage is permissible. For disposal of dead, they practice both burial and cremation methods. They are very much fond of songs and dances. Their annual festival is celebrated in April every year. Earlier they had well organized village council presided by the chief. Nowadays the village affairs are looked after by the panchayat. They follow the Tai script.

The Singpho villages are located on strategic positions. The house has an open balcony at the end where the ladies sit and spin, weave and embroider. The men folk tie their hair in a large knot on the crown of the head. Their dress is made up of coloured jacket of cotton. The chiefs sometimes love to dress up in the Burmese style. The outfit of the old consists of a piece of cotton cloth, often in large broad horizontal band of red and blue fastened round the waist, a jacket and a scarf. The men tattoo their limbs slightly and all the married women are tattooed on both legs from the ankle to the knee in broad parallel bands.

The Singphos know the technique of melting iron and manufacture *daos* with the help of very simple implements which are highly priced for their finish and durability (Chowdhury, 1979: 190). They are settled agriculturists and they practice both wet and dry cultivation.

14. The Galos

The Gallongs or Galos constitute one of the seventeen tribal groups inhabiting the West Siang District. They are the largest tribe inhabiting this district. They are concentrated in the circles named Along, Basar, Tirbin, Liromoba, Darak, Kaying and Likhabali. Agriculture is their main occupation. Jhum cultivation is still the prominent form although permanent cultivation is gaining momentum. Every special occasion is marked with sacrifice of animals. They believe in Donyi-Poloism.

The Galos have considered themselves as a separate tribe recently. Earlier they were considered a sub-tribe of Adi. S. Dutta Choudhury says, "Among the Adis, the Gallongs as a group are renowned and the most numerous" (1994: 71). Mopin the most popular festivals of the Adis, more particularly of the Galos in the month of April. Among the Galos, a family is composed of the father, mother, their sons with their wives and unmarried daughters as well as married daughters till they attain motherhood. The family is headed by the father and all the members recognize his authority. The sons inherit the property. The daughters in marriage are given a small dowry depending upon the amount of bride price received.

The marriage system of the Galos is remarkable in

many ways. Monogamy, polygamy and polyandry all in a complex form are in vogue in their society. The Galos have a sort of fraternal polyandry, whereby the unmarried young men in the family have a customary right to mate with the wives of their brothers. The high bride price does not usually permit each of the brothers in a family to have a wife for himself. The brother who married her by performing all the ceremonies is recognized as the actual husband. Cross cousin marriage is preferred. According to Choudhury,

> The marriageable age of Gallongs ranges from infancy to a little above adulthood. There is hardly any freedom in the selection of wives or husbands. The girls are not left unmarried till they can express their opinion. Cases of young girls not liking their husbands, to whom they were married in their childhood, are becoming numerous (1994: 101).

The Galos bury the dead. They prefer to site their villages on the high spurs of hills which may provide the villages a congenial living as well as security. The villages are generally of twenty to thirty houses on average. The houses are built on splits of bamboo and wooden planks are also used. There are few doors but hardly any window. The shape is square. The main part of the house is a big hall which serves as the bedroom, living room, and also dining room. A Galo house is decorated with horns and hides of the animals killed in hunting.

The male dress consists of a shirt called *lalik* with or without sleeves open in front, and a piece of loin cloth to cover the lower portion of the body. The skirt or *jesek* of the grown up Galo girls and women is used usually a white cloth locally woven, with black geometrical designs. The Galo use clothes of various designs and colours which they get from outside besides their own textiles. The hats of the Galos are fine artistic works in cane. The hat is very hard, water proof and strong enough to protect the head from the sun, rain, and storm.

15. The Mishmi

The Mishmis, living in the north and west of Lohit Valley were the first to come from Myanmar. They are short statured and have pronounced Mongoloid features.

The Mishmis live in long houses scattered over wide areas. A high bamboo palisade often surrounds a house and this becomes more evident when an epidemic breaks out. The palisade of spiked bamboo is supposed to keep evil spirits at bay.

Mishmis are divided into three main groups: Idus or Chulikatas, Digaru or Taraon and Miju or Kaman. The Idus live in the Dibang Valley District and the two other groups live in the Lohit District (Pandey, 1997:16). There are dialectical variations among them. Both men and women are inveterate smokers. They are jhum cultivators. Most of the works in the field are carried out by women and children. Maize and buckwheat are the staple crops - barley and wheat are also form the harvest in the top of the hill. They are expert in constructing bridge across the rivers and streams. The Mishmis visit the markets of Assam to sell musk, ivory, skin, and Mishmi *teeta* (a medicinal root) and purchase salt, cloth and other articles of use. The male dress of Jaroans and Kamans consists of a black or maroon colour coat and a carefully woven cane hat. There is an ornamental border on the coat. The Idus wear a kind of war coat made of nettle fibre, cotton and human hair strong enough to resist the thrust of an arrow. The dress and ornaments of the Mishmis are varied and beautiful. Idu Mishmi dresses are well designed and their sense of colour combination is admirable.

Some of their dances are Idu, Ah-ih, Mesala, Digaru, Beeiya and Nuiya. Ah-ih is basically a priest dance of Idu Mishmis. It depicts the action of a priest who, with his divine power tries to discover the cause of sickness of a person and seek solution for its cure. He performs the necessary rites and sacrifices. Mesala is a kind of death dance performed by Idu Mishmis when there is any death in tribal wars or fighting. Both the victor's and defeater's family perform the same ceremony with a view to succeeding either in taking revenge or to prevent any such revengeful action by the other party.

Reh (Idu Mishmis) and *Tamladu* (Taraon and Kaman Mishmis) are their important festivals. *Reh* is celebrated on the first of February. Animals and birds are sacrificed for the welfare and prosperity of the community. Idu dance is performed on the day. *Tamladu* is celebrated by offering prayers to God of Earth and Water for protection against natural calamity. The supreme - Lord Jebmalu, is worshipped for the prosperity and welfare of human beings, crops and domestic animals. The priest called Igu among the Idus performs a dance which has got special significance in their society (Pandey, 1997: 17).

In marriage, though monogamy is normal rule, polygamy is widely practiced. Tribe endogamy and clan exogamy is the fundamental rule. The system of payment of bride price is in vogue. They are agriculturists and expert weavers. They practice shifting cultivation. The Idus and Kaman (Mijus) have their own council of elders called Abbela and Pharai respectively, which are responsible for solving the disputes as per traditional customary laws. The existence of a super god, called Inni by Idus, Jab Mulu by Taraon and Matai by Mijus is believed by them. Inheritance of property always follows the male line. They bury the dead body and purificatory rites are performed with the help of priests.

16. The Tagin

The Tagins live in the interiors of Upper and Lower Subansiri Districts. Tagins are believed to have migrated from Penji, a village in Tibet. The house structure of the Tagins is peculiar in the sense that it is raised very high from the ground on the wooden and bamboo pillars. Staircases are used for reaching the first floor. The floor is made up of split bamboo. Mostly walls are made up of bamboo but in some cases, of wooden planks. There are several compartments. The house may be occupied by several families but each has separate compartment. They thus follow a joint family system like Nyshis. Each family is economically independent having land and other possessions but the senior most member is regarded as the head of the family. They stand united against raids and join hands for communal work, inter-clan feast or celebrations. Thus a Tagin house is a symbol of clan unity. They follow Jhum type of cultivation. They grow paddy, millet, maize, onion, brinjal, chilli, ginger etc.

Tagin society is divided into several classes. Nibus are a priestly class and known for their knowledge of myths, lore, power of negotiating with the spirits. They perform rites and sacrifices. Nites are the rich people who have got landed property, mithuns, pigs and ornaments and nyira or slave (Chowdhury, 1979: 142). Though monogamy is the rule, polygamy is practiced by the rich people who can afford to support several wives. Cross cousin marriages are not only preferred but also claimed as a right. Payment of rice is still in practice even now. Tribe endogamy and clan exogamy is the norm. The bride price is a must and may include a number of mithuns and cash (Das, 1995: 309). The disputes arising in a village or between the villages are settled by the council of elders. The Tagins believe in Si and Donyi (earth and sun) gods and also in existence of numerous spirits. Si-Donyi is the main annual festival, celebrated in the month of January to appease the creator (Pandey, 1997:23).

17. The Nah

Among the various tribes of the Upper Subansiri District Nah is a minor group inhabiting in the Taksing circle with the population of not less than eight hundred having eight villages as per Administrative report 1987-88 (Mitkong, 1999: 225). The Nahs are believed to have migrated from a Tibetan place called Lung near the Subansiri river. Nahs are Buddhist who perform the traditions of Lossar and Siju festivals. But in many different ways such as dialect, marriage system and many other cultural aspects they are similar to those of Nishis and Tagins. The role of village priest called Nyibu is significant specially for performing rituals and treatment of sickness.

The houses are constructed on raised platform and made of locally available bamboo and wood etc. The Nah society is patrilocal, patriarchal and patrilineal. The food habit is simple but delicious too. Upo and Chang are the common local beer made out of rice, millet (Pandey, 1997:24). Monogamy is the actual norm, but polygamy is permitted. Agriculture is their main livelihood and they practice jhum cultivation. The dress and costumes are simple which cover the whole body. They put more emphasis on the dress and ornaments of the head. The disputes are generally settled amicably by the elderly persons as per the traditional customary laws.

18. The Hill Miri

The Hill Miris live in the Lower Subansiri District. They do not at present preserve any racial memory as to the cause of their migration to their present home. They simply say that "they were made for the hills and appointed to dwell there" (Chowdhury, 1979: 97). Their villages are situated on the top or by the side of hills. The houses are built on raised platforms with locally available materials.

Their social structure consists of a number of Phraties comprising a group of clans. Besides the division of the Hill Miris into Phatries and clans, they have been divided into different broad groups according to popular conception. These are - Ghyghasi Miri, Ghasi Miri, Sarak Miri, the Panibotia and Tarbotia Miri. Both the men and the women of the different groups of the Hill Miris are found to dress differently. The women wear different types of ornaments made of silver, fine porcelain, beads of agate, cornelian and onyx as well as ordinary glass beads of all colours.

The family comprises of parents and unmarried children. In marriage both monogamy and polygamy is practiced. The tribe endogamy and clan exogamy is followed. The practice of bride price is prevalent. They bury their dead. The Miri village does not have a *morung* or bachelors' dormitory such as in the Adi village. There is a chief in each village who is responsible for maintaining safety in the village. They practice shifting cultivation. They raise crops like rice, millet, Indian corn, yam, sweet potatoes, small quantity of tobacco and red pepper. They believe in Si-Donyi representing the earth (si), the Sun (Donyi) (Pandey, 1997:23) and in the existence of numerous malevolent and benevolent gods and spirits. The Bori Boot is the main festival celebrated with enthusiasm and gaiety. The law and order of a village is looked after by a council of elders called Bango (ibid. 1997).

19. The Khamba

The Khambas inhabit the West Siang and the Upper Siang Districts. They are Buddhists people in and around the Tuting subdivision alongside the other Buddhist highlander - the Membas. It is believed that the Membas are the first group of people who emigrated from Tibet to this region about ten or eleven generations ago. Comparatively, the Khambas are recent immigrants who have migrated to this region from Tibet not more than five to six generations ago. Both these groups are closely akin in their language and customs and largely indistinguishable in appearance (Banerjee, 1999: 285). The Khamba is a blanket term being applied to anyone who came from Tibet. Strictly speaking Khambas are only those who hail from the Tibetan province of Kham. On the other hand, most of the Membas have migrated from Yamdok, Karmaling, Dakpa, Lhagykri and Kargpo areas in Tibet.

Their society is patrilocal, patriarchal and patrilineal and the family comprises of parents and unmarried children. In marriage tribe endogamy is maintained. Monogamy is the usual rule though polygamy is also practised. They build their houses with stones. Though agriculture is their main occupation, carpet weaving, wood caving, and domestication of the animals form an important means of livelihood. They perform masks and other dances during the celebration of the festivals and other social occasions. Influence of religion in every aspect of the lifestyle has been observed.

20. The Membas

The Membas are a numerically smaller tribe who are inhabitants of the Upper Siang District and profess Mahayana sect of Buddhism. Their life style is greatly influenced by religion. The family comprises of parents and unmarried children. Monogamy is the usual rule in marriage. They perform a number of panthamine dances on various occasions. Dhurba and Manne are the main festivals. The milk and milk products are important items of their food (Pandey, 1997:19). They are agriculturists and rearing of animals is one of their primary occupations. Their dress and ornaments are mainly woolen and well designed. The carpet making, wood carving etc. are their indigenous industries.

21. The Zhakring

The Zhakrings are a small group of people mentioned in the 1971 census having close racial and cultural affinities with the Tibetan people. The Zhakrings seem to have migrated from from Tibet and settled in the Dri Valley at Wallong Kibithoo circle of the Lohit District in Arunachal. They speak the Zakhring or Meyor dialect which is closely related to the Tibetan language and belongs to the greater Tibeto-Burman linguistic group (Barua, 1995: 406). At present they are found to inhabit the villages of Dong, Tinai, Kauhau, Mausai, Khaothi, Yakung, Saupkung, Upper Walong, Lower Walong etc. Their language has affiliation with the Bodic speech family. They have Mongoloid features. They follow a crude form of Buddhism. They are good cultivators and grow cereals on a large scale.

22. The Miji

The Mijis inhabit the West Kameng District, immediately to the North of the Akas. They are a small group of people differing very little from the Akas. The Mijis call themselves Dammai in their own language. They believe that their ancestor was the Sajolang who descended from the sky through a silver ladder along with the forefather of the Tibetans and other tribes. They are scattered over twenty five villages. They are almost exclusively jhum cultivators. They raise two crops in the same year.

The Miji society appears to be vertically divided into two classes - Nyubbu and Nyullu, which are again subdivided into several subgroups - Sangcha-zu, Kimzu-zu and Phangdang-zu. Nyullos are treated as social inferiors. They have to carry loads for Nyubbus. Separate utensils are used for cooking food as the Nyubbus are forbidden from taking the head, liver and legs of the slaughtered cow or pig (Chowdhury, 1979: 75-76). The system of slavery was in existence among the Mijis. The houses are made on a raised platform. They generally practice polygamy. The system of bride price is prevalent. They bury the dead. There are separate burial grounds for Nyubbus and Nyullus. They rear mithuns, pigs, goats, and fowls. The mithuns are used either as a bride price or as a sacrifice to be offered in a feast.

The Khan is the annual festival celebrated in the month of March. Their religious belief may be termed animistic as the idea of spirits is predominant. Jang Lang Nui is considered as the most powerful of all the spirits. To appease the spirits they perform rituals with the help of priests.

23. The Sulung (Puroik)

Another small and distinct group scattered over the

hilly terrains of the East Kameng and Lower Subansiri Districts are the Sulungs. They are considered as one of the most primitive racial group inhabiting in the area (Deuri, 1982: 107). At one time they lived in a semi-nomadic manner, roaming about in the forests in search of wild games and jungle produce. They were good blacksmiths and brass founders.

The village of the Sulung comprise of a few houses only. They prefer nuclear family. They may be called as a migratory tribe, because earlier they were food gatherers and they frequently shifted their villages where a local variety of palm tree, *sago*, was available. Tase (Sago Palm) is their staple food (Chowdhury, 1979:112). The extracting of flour from the tree is an elaborate process. They now practice jhum cultivation. The dress and ornaments are very simple type. The women tattoo their faces with their local ingredients. In marriage, tribe endogamy and clan exogamy is strictly followed. Monogamy is the normal rule and they prefer cross-cousin marriage. Bride price is prevalent.

All the disputes and other affairs of the village are settled amicably by the village elders as they do not have a well-organized village council. Slavery was prevalent among them and all their socio-economic or political problems were settled by their masters. The common method of disposal of dead is by burial. A palisade is built around the place and it is customary to hang on a post the personal belongings of the dead inside the enclosure.

Some interesting food taboos have been noted among the Sulungs. Animals killed during the hunting expeditions should not be eaten by a man's mother, sister and in fact, any near female relatives, excepting his own daughter. There is also said to be a taboo on the eating of domestic fowls by young girls. They believe in a number of malevolent and benevolent spirits and deities. The rituals and ceremonies are performed with the help of priest.

In spite of their close proximity to the Nyishis, they

preserve their language which is said to be totally different from any language either by the eastern or western branch of the Nyshi tribe. They have a distinct language of their own. But the Sulungs claim kinship with the Khowas or Buguns. The language spoken both by the Khowas and the Sulungs are mutually understandable.

24. The Lishpa and Chugpa

Within a radius of four or five miles of Dirang, two very small groups of people are found. They are called Lishpa and Chugpa. In many respects they are akin to the Monpas, having a similar culture and religion (Chowdhury, 1979: 46). Strangely enough, they speak a language, which is completely different and distinct from the Monpa language. The Monpas consider the Lishpa and Chugpa people as socially inferior to themselves. And they are unwilling to give away their daughters in marriage to them. It is commonly believed that they are descended from an early wave that migrated from Bhutan (ibid. 1979).

25. The Rishi-Mashi

Towards further north, occupying areas between the Kamla and the Subansiri which actually form the southern face of a 3658 meter range, there is a group called Rishi-Mashi. They are said to have "mongoloid features of a more delicate and progressive type" (ibid. 1979: 106).

26. The Nidu-Mora

A tribal group, having marital relation with the Rishi-Mashi but maintaining a separate identity, is known as Nidu-Mora. This group is located in the Upper Subansiri valley (ibid. 1979).

2

The Thematic Concerns of Lummer Dai and Yeshe Dorjee Thongchi

Section A

Lummer Dai is one of the prominent writers of Arunachal Pradesh who writes in Assamese. In his novels the tribal society of Arunachal is presented vividly. He wrote five novels, the last of which he finished writing just seven days before his death:

(i) *P aharar Xile Xile* (In the Midst of the Rocks of the Hill) (1961)

(ii) *Prithivir Hanhi* (The Smile of the Earth) (1963)

(iii) *Mon Aru Mon* (Heart to Heart) (1968)

(v) *Kanyar Mulya* (Bride Price) (1978) and

(v) *Upar Mahal* (Higher Level) (2002)

Here is an attempt to analyse the general thematic concerns of the novels of the writer. Almost all his novels (except the last one *Upar Mahal*) are somehow related to the traditional social life of the Adis. Dai depicts the rituals, beliefs and social practices of the Adi community in his novels. In fact, an insider's view of the life and community of the Adis can be seen through his novels. Having said this, I am aware of the fact what Kirin Narayan says of anthropologists might also be true to some extent in the case of literary writers:

> Factors such as education, gender, sexual orientation, class, race, or sheer duration of contacts may at different times outweigh the cultural identity we associate with insider or outsider status. (1993: 671-672)

Similarly, Rosaldo (1989) opines that every anthropologist (and by extension, I argue, every writer of ethnographic fiction) exhibits a "multiplex subjectivity" with many crosscutting identifications:

> Cultural identities come from somewhere, have histories. But like everything which is historical, they [identities] undergo constant trans-formations. Far from being eternally fixed in some essentialised past, they are subject to the continuous "play" of history, culture and power (1989: 70).

Each of Lummer Dai's novels can be examined from various perspectives as these throw light on different themes. The themes of love and betrayal, pain and conflict (both internal and external), of social customs, of conflict between generations as well as of social transformation taking place in the modern society of Arunachal - all these are the concerns of Lummer Dai in his novels.

Paharar Xile Xile (1961), Lummer Dai's first novel, is considered to be the first novel written about the tribal people of Arunachal by an Arunachalee. Though in the 'Preface' or 'Introduction' of the novel Dai compares his writing to the "unsavoury cuisine of a new bride", Birendra Kumar Bhattacharya says that in this novel we are introduced to the cruelty that human beings are capable of. For him, this novel is a realistic picture of a girl – Bati.

At the very outset of the book, Lummer Dai says a few words which provide a clue in interpreting the thematic concerns of the book. Bati, the main character of the novel is compared to a wild flower which, though not nurtured by anyone, spreads its smell all around. He says that women like Bati who are utterly deprived of the advantages of modern civilization, are not rare in society. Dai attempts to show not only the position of woman in society but also the love, suffering, and anguish that she experiences. Moreover, the practice of 'slavery' is another important theme of the novel. In a sense the story of the novel revolves round this theme. The writer mentions in the preface that though 'slavery' has been abolished in Adi society, one who is born to a slave woman is still considered to belong to the lower class. There are impediments in the way of inter-class marriages in society. The novel is set against rural life and therefore, it is full of traditional beliefs and social customs. There is also a sense of protest against the prevalent evils of society along with the clash between two generations.

Though Bati is in love with Jalam she does not accept his love for her as she is quite sure about the implication of belonging to a family of erstwhile slaves. Jalam is the son of the *Gaonburha* (Village-head/chief) and thus obviously belongs to the upper-class. Therefore, Bati refuses Jalam's proposal time and again. The writer has portrayed Bati as the typical woman from a patriarchal society who can not even freely express her love. Once when Jalam insists that they should get married, she says:

Why should I marry you if your parents won't take anything prepared and served by me? My responsibility as a daughter-in-law would be to look after my in-laws and serve my husband. If that is not possible, there is no sense in getting married. (*Paharar Xile Xile*: 6)

One of the important themes of the novel is the exploitation of women in a male-dominated society. When Bati is accused of theft, her own husband, who was once madly in love with her, drags her out of the house with the kind of cruelty that is compared by the novelist to the indifference with which one discards the corpse of a dog. At that time Bati is seriously ill. However, even at this moment she behaves like a typical suppressed woman from a male-dominated society. She does not protest or cry. Even at this moment, she thinks of how to serve her husband and brother-in-law, her family members (*Paharar Xile Xile:* 77). She keeps thinking about her husband and in-laws till her last breath and tells her mother: "Send forty *agin* (a measurement of Adis) of rice to Jalam" (*Paharar Xile Xile:* 79). When her parents come to take her home she protests:

> I did not come here to go back to your house. My husband's house is my only shelter. If I have to die, I will die here. Even though you gave me birth, my husband is closer to me than you (*Paharar Xile Xile:* 78).

Paharar Xile Xile is a novel about love and separation, jealousy and torture (both physical and mental), and above all, a story of a young couple who have to suffer a lot due to the cruel system of slavery. The novel was written in the fifties and, depicts the Arunachalee society of the time when slavery was still prevalent. The reference to slavery is first heard in Bati's mouth when she refuses Jalam's proposal:

> If you marry me, your parents will be very upset as I am the daughter of a slave (*Paharar Xile Xile:* 6).

References to the practice of slavery occur all through the novel. Jalam's father, Bijong gets angry when he comes to hear about his son's relationship with Bati and exclaims in anger:

> I do not want a slave's daughter as my daughter-in-law. I do not want a single drop of slave's blood in my progeny (*Paharar Xile Xile:* 16).

All said and done, one has to accept that Bati has to suffer such torture and pain not only because of her identity as a woman but because of her identity as a slave's daughter which strengthens the issue of slavery as a major thematic concern of the novel.

Lummer Dai depicts the traditional customs and rituals of Adi society in *Paharar Xile Xile*. He is capable of assimilating the characters with the plot of the novel. The plot progresses very smoothly with the description of different Adi rituals and customs. In the rural society that Dai depicts, the girl is brought up with a clear set of dos and don'ts. The writer portrays Bati as a girl living in rural Arunachal and assigns her some works which are typical of a rural Adi girl. There is a description of traditional Adi women in the very first page of the novel:

> Some mothers used to breast-feed their babies while carrying firewood in the 'horas' (*Paharar Xile Xile:* 1).

Dai tries to show some traditional beliefs of the Adi society through the novel. The Adis have a belief that when someone in the family dies, then a new handwoven *jim* cloth (worn like a blanket) should be kept on the graveyard of the dead. This becomes clear in the course of day-to-day conversations among the women folk, especially when one of them talks about the rainy days due to which they could not spin thread:

> "If we were to lose someone from the family, there is not a single piece of new jim cloth to keep on the graveyard" (*Paharar Xile Xile:* 4).

Another ritual of traditional Adi society is that one of the groom's sisters should be married to someone in the bride's family. In *Paharar Xile Xile,* Jadi is supposed to be married to Jalam as his sister was married to her brother. When Bati turns down Jalam's proposal, she says:

> There is a girl whom you have earned in exchange of your sister (*Paharar Xile Xile:* 5-6).

Dreams are believed in and interpreted in a specific way in the Adi society. The Adis believe that if they are unsuccessful in climbing mountains and crossing rivers in dreams, they have to face a lot of problems and difficulties. When Jalam sees such a dream he thinks:

> He could understand quite well that even if he made thousands of attempts to marry Bati, it will all be in vain (*Paharar Xile Xile:* 30)

Every society is unique in its performance of ritual regarding birth and death and even marriage. In a traditional Adi society after the engagement of the bride and the groom, the bride would continue to live in her parents' house. She would go to her in-laws' after two to three years. The writer introduces us to this custom when Jalam refuses to allow his bride to stay with her parents after she is engaged to be married to him (*Paharar Xile Xile:* 49). Another custom related to death is that a husband who loses his wife is not supposed to perform any household activities (*Paharar Xile Xile:* 48). In the Adi society, there is a tradition of biting the tooth of a dead tiger to prove one's innocence. This is done by Bati when she is accused of stealing rice.

Besides, the novel is replete with words and ideas associated with traditional Adi society. The ways of welcoming guests with *apong*; the mention of Aran (*Paharar Xile Xile:* 14) - the Adi festival; the *mosup* and the young men from the *mosup* and their power; mention of *hora* (*Paharar Xile Xile:* 1), *kebang* and *agin* (*Paharar Xile Xile:* 61); eating habits with maize, potatoes and tubers - all build up the image of a traditional Adi society.

Mosup is a platform for the young man of the village where they spend their nights. The young men living in the *mosup* are known as *mosup deka*. The village-head (Gaonburha) would send such *mosup dekas* to other places as messengers. Besides the *mosup dekas* would guard the village from enemies. (*Paharar Xile Xile:* 28). The *mosup* has certain laws which are to be obeyed by each and every member. If a defaulter is found, he is fined in cash as well as in kind (pigs or mithuns) according to the gravity of the crime. We come to know about this when the leader of the *mosup* tells Jalam:

> We have taken the largest pig from your sty (*Paharar Xile Xile:* 36).

And again,

> You have to give Rs. 50 and a pig to the *mosup* as you have not obeyed the *mosup* and its rules (*Paharar Xile Xile:* 32)

Jalam was absent from the *mosup* without informing others due to which he is fined. In case of serious offences, the houses of the guilty are raided. This happens when Bati is accused of burning the rice field and there are "evidences" that go against her. Pigs and mithuns are forcefully taken away several times from Jalam's pig-sty for disobeying the *mosup* as well as for the charges brought against Bati.

The novel has a rich variety of thematic concerns including love and separation and the destructive power of jealousy. The writer is successful in depicting two lovers behaving with spontaneity and sincerity. The metaphor of 'an oasis in a desert' that the writer uses to compare Bati's love for Jalam is worthwhile in the context. Their pathetic separation, Bati's death and Jadi's jealously which leads to her insanity, reflect the writer's simple sorrow-striken heart which could not accept Jadi as a "normal" girl after all the misdeeds that she had committed.

Prithivir Hanhi (1963) begins with a vivid description of a *mosup* and an Adi village. This novel, like the other novels we have discussed, has many references to Adi social life, including the Etoro and Solung festivals, *delong* dance and *kebang*. The *delong* dance is performed in the Etoro festival on a full moon night. This is a dance meant for boys. In the Etoro festival the boys dance till midnight dressed in girl's attire and ornaments. Similarly, young girls dress up like boys. There is a vivid description of it in the novel *Prithivir Hanhi*:

> Liyi, Gasi, and many other girls are in search of friends. They are also collecting ornaments of girls. They will look at the handsome boys wonderfully dressed in girl's attire and ornaments (*Prithivir Hanhi*: 93-94).

There are descriptions of the *ponung* dance, i.e., the dance of girls dressed in male clothes. The Etoro festival is celebrated in the *mosup* of the village and the *miri* (priest) initiates it by chanting mantras. He sings the myth of origin of the world and its creatures and the youths of the *mosup* revolve round it and then dance together. Dai has elaborately described the Etoro festival and the rituals related to it in this novel. There is a long reference to a song sung in the *delong* dance (*Prithivir Hanhi*: 97-98). Even as the story progresses smoothly, there is mention of varied rituals, beliefs and practices of Adi society. In the first paragraph of the novel the protagonist Karduk is seen sleeping in a *mosup*. The mention of *jim* clothes (*Prithivir Hanhi*: 2) and the references later on to the selling of *jim* clothes in the market of Diburo show another tradition of the Adis. The Adis used to sell their *jim* clothes in the markets of Assam and there is a long and vivid description of this process in the novel. This tradition of selling *jim* clothes is not just a mere super-addition but has a lot to do with the story of the novel. This helps the readers to delve deep into the complexities of the human mind. Moreover, this whole business of selling *jim* clothes helps the readers to know the character of Kardug better who is not presented as an out and out virtuous character by the novelist.

The novel presents a history of the Adi society. It contains a comprehensive description of all the minute details of Adi society (Borpujari, 1999: 55).

The novel also shows the transformation taking place in the society and raises questions regarding the authenticity of *kebang* and people's protest against the unfairness that has crept into it. Satyendra Nath Sharma says, "The novelist does not shy away from exposing the faults in *kebang* or criticizing it" (1976: 245). In almost all the novels by Lummer Dai, the *kebang* plays an important role. The *kebang* as the highest judicial institution of the Adis, plays a role of immense significance in the Adi society. Liyi protests against the *kebang's* decision and openly speaks out against Kardug at the end of the novel. Liyi being physically assaulted by Kardug in the *kebang* itself is something which clearly shows the lenience of the *kebang* and the transformation taking place in society due to the loss of faith of the new generation in the traditional institutions. In response to Gasi's querry "Are you late?" Bangkong answers, "Yes, but it is not a serious matter. Now-a-days the rules of the *mosup* are no longer as stern as before. There are no more any hard and fast rules" (*Prithivir Hanhi*: 107).

The superstitious beliefs of the tribal people are referred to in the novel. In the second page of the novel there is a reference to the superstition associated with bats:

> The bat foretold some impending misfortune. The bat is a very inauspicious animal (*Prithivir Hanhi*: 2)

When Kardug goes out in search of his Mithuns, he comes across a snake and thinks, "It is ominous to see a snake in the forest. Then he cut off a branch of a tree and left it at the place where he had seen the snake and after spitting said, "Go away! I hope I will not see you again" (*Prithivir Hanhi*: 13).

The Adis' belief in supernatural beings is also seen in the words of Kardug's mother when she says, "Don't say that your wound is slight. If a spirit happens to hear that, it will swell up the wound" (*Prithivir Hanhi*: 16). When Kardug, Lib and Bangkong go out for hunting and sleep under a fig tree, Lib expresses his fear of *ban devata* (the spirit of the forest). He believes that there are spirits in the fig tree (*Prithivir Hanhi*: 70). When Lib is able to hunt a deer for the first time, Kardug's mother tells him:

You should offer worship to the *Ban devata*. He has showered his blessings on you. This happens very rarely. If you can satisfy *ban devata* by offering worship, you'll be an expert hunter (*Prithivir Hanhi*: 91).

Lib, a traditional Adi boy, has deep faith in all such supernatural entities. When he describes the way Siram, one of his pigs, recuperates from very serious sickness, he says, "I offered some rice in the name of the god of swine ..." (*Prithivir Hanhi*: 124).Lib promises to god that he would sacrifice Siram in the coming Dorung festival if the swine is saved.

Prithivir Hanhi is about love, friendship, jealousy and cruelty. As discussed in the context of the other novels of Lummer Dai, there is mention of hunting, search for mithuns, the process of identification of mithuns by looking at the cut in the ear, gathering of firewood by Adi woman in *horas*, yarning of clothes, grinding rice in the *ural* etc. According to the tradition of the Adis, the bride and groom sleep together after engagement in the bride's place: "Gasi thought: "Kardug will sleep here from today". She used to share the same mat with her sister at night" (*Prithivir Hanhi*: 35). It is considered to be a matter of shame if a young man gets up late in the bride's place (before marriage):

> During the first few days of one's sojourn in the bride's place, one should be seen neither entering nor leaving the bride's place. If seen, he would be ridiculed as being lazy. One has to enter the house after everyone goes to sleep and has to leave before dawn (*Prithivir Hanhi*: 46).

The tradition of *ape* (the custom of marriage of the husband's sister with the wife's brother) (*Prithivir Hanhi*: 54/55) and the name of a son beginning with the last two syllable of his father's name (Kardug's son is named Dugbang) (*Prithivir Hanhi*: 56) are some of the Adi practices that the author uses with the progress of the novel. *Mousup*, the centre of the rural life of the Adis is discussed in detail in *Prithivir Hanhi*. The youth have to spend their night in the *mosup* till their marriage. They have to strictly follow the ritulas of the *mosup*. Moreover they are accountable for their inability to be present in the *mosup* in time.

Tilottama Misra says:

> Lummer Dai's approach is often ruthlessly realistic and critical when he lays bare some of the social tendencies that have brought unhappiness to the lives of the Adi people, especially to the women (2001: 34).

Like other novels of Lummer Dai, *Prithivir Hanhi* shows the novelist's concern for the status of women in society. Kardug is a male chauvinist who holds women in low esteem. Initially he falls for Gasi and then as he feels drawn toward Liyi, he accuses Gasi of having an affair with Bangkong. Not only does Kardug deceive Gasi but even lends Liyi a fatal blow in the very presence of the *kebang*.

Kardug's chauvinist mind which makes him think of woman as a mere commodity is reflected in the following sentence:

> I'm leaving you to Bangkong ... Just sitting at home I'll get Rs. 200 in cash or a mithun from your new husband as compensation. How many people have the fortune of acquiring such a lot of property just like that without working hard? (*Prithivir Hanhi*: 114)

Woman is seen as a commodity that can be exchanged for a mithun or some cash.

The Ministry of Education and Youth Affairs, Government of India honoured Lummer Dai in 1970 for his novel *Mon aru Mon* (Heart to Heart) with the Talent Search award. This novel was published in 1968. The story of the novel is about an old widow and her pet dog Bomong. The old widow comes to have a very close relation with the dog after the passing away of all her near and dear ones. The dog is no less than a loving relative to grandmother Gidum. *Mon aru Mon* is really a wonderful addition to the field of Assamese literature. A dog as one of the central characters of a novel is a brilliant addition to the Assamese novel.

The major thematic concern of *Mon aru Mon* is the deep unselfish love of Gidum for a dog that she does not consider to be just an animal but a being with human qualities and emotions. In fact, at the end of the novel we are left wondering whether Gidum or the Bomong, the dog, is the real protagonist of the novel. The novel talks more about the dog, its activities and behaviour, its loyalty and finally, its death followed by Gidum's death. The novel begins with a long lecture by Gidum and obviously it is addressed to the dog. What is amazing here is that the way in which she talks to the dog does not seem to be a conversation with an animal that is incapable of verbal communicaton. Here, the reader is introduced to the bond of love and affection between Gidum and her pet. When there is a scarcity of rice, Gidum offers her own share of rice to Bomong while she herself has tubers. In short, Bomong becomes an inseparable part of her world which is small but contented.

A major thematic concern of the novel *Mon aru Mon* is the customs, beliefs and superstitions of the Adi society that the novelist tries to depict. In fact, in this novel Dai talks about a variety of folk beliefs prevalent in the traditional Adi society. These folk beliefs are either described by the narrator or observed by other character(s). The beliefs associated with the *ari* (a still born child or a child that dies immediately after birth) is being introduced by the author to the readers at the beginning of the novel. The *gam* (chief) of the village declares, "No one can go outside the village" (*Mon aru Mon*: 2). When Gidum asks about what has happened, she is told that Dadam has given birth to a still-born child. And Gidum automatically infers that no one is supposed to leave the village or engage oneself in any other work as it might harm him/her. Following the same belief, Gidum scolds some young girls for washing clothes during this period following the birth of the still-born child. Similarly, Dai introduces us to another such traditional belief in this context - if someone spins yarn at a time when someone else from the same village is bitten by a tiger, he/she has to face dire consequences. Therefore, Gidum says:

Now I remember that she was spinning yarn at the time when Gamuk was bitten by a tiger. How would she be able to evade the penetrating gaze of the spirit? She has to face all these hardships due to her lack of regard for social norms (*Mon aru Mon*: 3).

Gidum discovers a parallel between Dadam's misfortune and her act of spinning yarn when a tiger happened to bite Gamuk.

The novel highlights a variety of folk beliefs associated with the Adi people's day to day activities, diseases and death. An example of this is the belief in the goddess Ngippong. Verrier Elwin in his *Myths of the North East Frontier of India* talks about this belief and says, "The Ngippong *wiyus* kill women in childbirth" (1958: 302). Dai too explains the concept in the footnote as:

> If a woman dies when she is pregnant or in labour pain or at the time of delivery, she is called Ngippong. And the goddess who is responsible for such death is believed to be very cruel and spiteful and the belief is that the goddess enters the body of the woman (*Mon aru Mon*: 8).

Gidum firmly believes that the goddess Ngippong has caused Dadam's pain by entering her body and thus ultimately leads to her death. Gidum even tries to chase Ngippong away from Dadam's body by beating her with a bunch of leaves that causes itching sensation. There is mention of another belief associated with the call of an owl. In the Adi society people believe that this is a bad omen signalling the impending death of someone. Therefore, when she hears the call of an owl, she says:

Just shut up! It's a futile attempt. You can't do anything till I am around (*Mon aru Mon*: 11).

Donyi-Dapong is another folk belief of the Adis that finds mention in this novel and Dai has explained its meaning in the footnote. It is a form of worship of the sun. A white hen is sacrificed as it is considered the symbol of sunlight (*Mon aru Mon*: 36). Therefore, when it rains heavily, Gidum chides the rain and says:

> Wait till I ask the villagers to offer Donyi-Dapong puja tomorrow and you wither in the heat of the sun (*Mon aru Mon*: 36).

Besides these, storms and heavy rains are considered as the symbol of impending misfortune (*Mon aru Mon*: 65). Similarly, the dream of a rainbow is believed to be the prediction of someone's death (*Mon aru Mon*: 135-36). Therefore, when Gidum talks about seeing a rainbow in her dream, Badang's mother says:

> It is really a very bad dream. Someone is going to die a horrible death (*Mon aru Mon*: 136).

The Adis believe that beating against the wall with spears in a house where someone has met with an unnatural death helps to chase away the evil spirits. Therefore when Gamling dies after being bitten by a dog, the villagers start to beat against the wall of Gamling's house with the help of spears (*Mon aru Mon*: 86).

Gidum's deep faith in the need to observe certain practices and rituals includes offering worship during the Solung festival for good harvest (*Mon aru Mon*: 1), recovery from bouts of fever (*Mon aru Mon*: 59), examining the human body through eggs and rice (*Mon aru Mon*: 92) and sacrificing birds and animals for the well-being of people. Similarly, the mention of *koson puja* in favour of *bandevata*, sacrificing a hen in a small *koson*, and a mithun in a big *koson* (*Mon aru Mon*: 123 - 124) and the vivid description of ways of sacrifice take us to the world of Adi beliefs and rituals. *Dotgang* (*Mon aru Mon*: 150) is a way of worship by sacrificing pig or mithun in favour of the dead; it is believed that this pig or mithun becomes the servant of the soul of the dead.

An important concern of the novel is that there is a conflict - a clash between characters, between two generations and between tradition and modernity. It has been already mentioned that this novel highlights a variety of rituals, customs and traditional beliefs. It is important to note that the conflict in the novel originates from differences of opinion regarding observance of such beliefs.

There are two sets of characters as such - one that strictly adheres to the old customs, traditions and beliefs and the other that strives for a change, a step towards modernity and rational thinking. The title of the novel is self-explanatory in so far as the theme is concerned. The Assamese equivalent for *Mon aru Mon* is 'Different Minds'. The title refers to Gidum's superstitious mind and Badang's 'rational' mind and the conflict between the two. Jeetanjali Barpujari rightly says:

> The novel bears the stamp of the conflict in the mind between reason and superstition. On the one hand, there is Badang's rational, modern and progressive mind that is free of superstition and on the other hand, there is Gidum's superstitious traditional mind (1999: 60).

The clash between these two minds is clearly visible all through the novel. We come to know about Gidum's obstinacy from her observance of the traditional rituals. She sacrifices a hen in *koson puja* for quick recovery from fever. And when she does recover finally, she thinks that it is because of the *puja*. At the same time she restricts herself from going to the field after that puja for three days, though she has a lot of work there. This is because she has respect and faith in the age-old belief that one cannot work in the field for three days following the *koson puja*.

Badang stands for the progressive rational youth who

somehow convinces Gidum to take medicine to recover from fever. He never believes in such superstitions and therefore, can go out of his village even after seeing a storm in dream (which is considered ominous). Borpujari rightly says, "Lummer Dai is succesful in depicting a clear picture of eternal conflict between the new and the old" (1999: 61).

Related to the theme of the conflict of two generations is the note of reformation or change. We sense the desire for change in Badang's mind when he overcomes the struggle within his mind and proceeds to Pasighat the day after he has seen a storm in a dream (*Mon aru Mon*: 134-35). Both Badang and Ajong being literate representatives of modern progressive society believe in something new and positve. Ajong says,

> Everything will be all right in due time. Not just a single grandmother Gidum but thousands of such grandmothers will be compelled to welcome the new, reformed world in place of the new (*Mon aru Mon*: 116).

The education and experience of the outer world enables Badang to think against the social norms of his soicety:

> Grandmother Gidum will not be able to go to the field for three days as she has performed *koson puja*. She firmly believes that the *bandevata* will be dissatisfied if she does so. Thus the valuable time of a lot of people is wasted (*Mon aru Mon*: 123).

Here, we sense the reformative zeal and the urge for change in Badang who being an upcoming new generation youth looks for the betterment of his society.

Lummer Dai received the Sitanath Brahma Choudhuri award from the Assam Sahitya Sabha for *Kanyar Mulya* (1978) which is his most famous novel. It depicts the Adi social life through its presentation of one of the oldest social institutions of human society i.e. marriage. The protagonist Gumba is the victim of the long-established tradition of the Adis, - that is the social practice of taking and giving bride price. In spite of being almost dead, she does not give up her protest against the long-established system of child marriage. Kargum, her father, is of course a representative of the old generation who strictly adheres to the age old traditions, social practices and beliefs but dwells in conflict between his love, care and sense of duty for his daughter and his sense of responsibility toward the society. When examined thoroughly, *Kanyar Mulya* seems to enclose varieties of different ideas and themes that the writer tries to focus on. From the title itself, it is clear that the novel is about bride price which was a prevalent social custom of the Adi community. I was informed by Prof. Tamo Mibang, an Adi intellectual, in a personal interaction that the system of bride price is still prevalent in the Adi villages, and though at one point of time child marriages were prevalent amongst the affluent families, they have ceased to exist now. However, according to the social custom of the Adis as shown in Dai's novel, the marriage of a girl is fixed at an early age, sometimes just after birth, taking the price of the bride from the groom's family. In *Kanyar Mulya,* when Gumba is still a little child, her father Kargum enters into an agreement with Mindak according to which after a certain period of time Gumba will have to go to Mindak's house as his daughter-in-law (as the wife of his son, Dakat). As she has grown up and studied in the high schools in town, she becomes aware of the demerits of social practices such as child marriage and bride price. As such, she starts to protest against the social order. Even before she comes to know about her own marriage, she voices her strong feelings against child marriage and bride price while talking to her friend:

> Men consider woman as commodity. Immediately after a female baby is born, they think that it is the acquisition of a new property. This custom of

> selling and buying of girls must be done away with very soon (*Kanyar Mulya*: 8).

Gumba's marriage is fixed at such a tender age when it is impossible for one to even comprehend gendered division in society. When she learns about her marriage, she protests and the pain that she experiences is obvious. She cries and says

> What do I hear? Who is my husband? Does it mean that you have sold me silently? (*Kanyar Mulya*: 9)

She protests against her father:

> Kill me father. Do kill me. It would be better to kill than to sell me (*Kanyar Mulya*: 9).

She does not hesitate to ask her parents:

> Did you give birth to me just to sell? Then why did you give birth to me? Why? (*Kanyar Mulya*: 9)

Even when she is forcefully taken from school to her in-laws' by Mindak and his relatives, she puts up a bold show against the old practice:

> The days are gone when you could sell and buy human beings. Today there is no girl who will allow herself to be sold and bought like mithuns and cows (*Kanyar Mulya*: 37).

Even when she is kept in Mindak's house, Gumba does not give up her protest. In fact, she is determined not to surrender till death. She refuses to eat anything for sixteen days (*Kanyar Mulya*: 67) when she is forcefully kept in Mindak's house. She tells Dakbi (Mindak's sister): "I am fighting against acts of inhumanity and babarism that are committed against women" (*Kanyar Mulya*: 65).

There is a distinct note of feminism in the novel. Dai has shown the traditional patriarchal society of the Adis as it is, without any attempt to either scoff at it or suggest remedies for the maladies. Feminism is manifested in the novel in the form of a protest against the established patriarchal norms and the attempt to establish one's own identity. In fact, the entire novel is about Gumba's physical and mental torture and her strong protest against it. Patriarchal society considers woman to be a commodity, a thing which can be possessed by any means. The girl is just like a commodity which is not expected to have its own mind, likes and dislikes, not even the right to protest:

> No one can raise an objection if we bury you alive; none, not even your father. It does not matter what we do to you, you don't have the right to protest (*Kanyar Mulya*: 68).

His patriarchal ego is evident when Joken proudly announces in front of Gumba's father:

> After taking the price of Gumba, you do not have any right over her. She is ours we can do with her whatever we want. Even if she is buried alive, no one can say anything (*Kanyar Mulya*: 72).

Again in the Kebang (the meeting of the village elders) Tumde says:

> We have bought you. So we need not argue with you about anything. We can do whatever we want with you. After all, who would bother to argue with a purchased commodity? (*Kanyar Mulya*: 91)

But in the same context, Gumba protests and says:

> You cannot buy a human being with your five Mithuns, seven cows and a few dishes and bowls (*Kanyar Mulya*: 91).

This is an expression of revolt against the traditional patriarchal society. Again when the Gam (Chief) once says, "If we, the leaders of society start breaking social norms, our forefathers will not forgive us" (*Kanyar Mulya*: 94). Gumba proclaims:

> Kill me and preserve your social norm (*Kanyar Mulya*: 94)

In the same *kebang* another Head Chief Itum says,

> Even if Gumba dies without eating anything in Mindak's house, we need not worry. Mindak has bought her and only he has right over her. Despite being her father Kargum has also lost his right over her once he has taken her bride price .... (*Kanyar Mulya*: 98).

Gumba asserts her identity by saying: "I am the only one fit to talk about me. No one else can" (*Kanyar Mulya*: 41).

Even Yabi, Mindak's daughter once wants to say:

> Woman is not a play thing that what ever you wish, you can do with it. The female body is not a commodity to exhibit for others to look at (*Kanyar Mulya*: 43).

Conflict of two generations as well as the transformation taking place in the society is another recurring theme of *Kanyar Mulya*. Minjum, who used to write letters for Kargum and is a representative of the modern society, says:

> No one can stop the flow of water. In the same way, change in the society is inevitable; no one can prevent change. Earlier we used to torture women, we did not treat them humanely, but like cows and as commodity..... but those days are gone. The women have understood that they are human beings. Therefore, they are revolting against the torture of men (*Kanyar Mulya*: 76).

FInally Kargum understands and accepts his words and starts believing that he was wrong and proclaims:

> I cannot allow my daughter to be sacrificed for the sake of social norms (*Kanyar Mulya*: 77).

The transformation that takes place in the society is seen in the ultimate judgement of the Kebang when it declares that Gumba should not go to Mindak's house. By accepting to take back the bride price already paid by Kargum, Mindak too succumbs to the force of social change and transformation. Social norm and old practices can be broken for the sake of humanity. For the sake of a human life old rules can be discarded, this is what is established at the end of the novel which clearly shows the transformation taking place in the society.

The Kebang where Gumba's case is discussed is not only a trial but a battle ground between the old and the new. It reflects the conflict between age old beliefs and new realization. Minjum's long lecture depicts the change that has come about in the attitude of the new generation:

> The problem is not regarding the return of the bride price. We have a problem with child-marriage and forced-marriage. We, the men have been torturing the souls of countless women. But still we are not contented. When a baby girl is born we think that she will add to our property ... During the present period of crisis in the bride market, we do not shy away from buying a daughter-in-law just for a few mithuns and cows even when she is still in the womb of the mother. (*Kanyar Mulya*: 94)

This shows the social transformation taking place in society.

Gaken, the head chief, also talks about social change:

> The Adi rules and rituals have taken thousand different forms from the days of Abotani ... People do not eat the kind of food that they used to eat thousand years ago. The dress we wear, the language we speak today was not worn or spoken by our ancestors ... We have been firsthand witnesses of changes in social practices and beliefs (*Kanyar Mulya*: 99).

All these discussions clearly show the transformation taking place not only in the society but in the minds of the people too.

The novel reveals the inner conflict of a particular character. Kargum, Gumba's father, is seen to be quite overwhelmed by some situations and his words clearly express his mental agony and inner conflict. In the very first page of the novel, readers are introduced to the mental tension and inner conflict of Kargum. When thinking about Gumba, he says,

> She is old enough to go to her in-laws' place. She should be sent there. What will she do in school? She will never be an officer. But what if she can become one? (*Kanyar Mulya*: 1)

When Gumba is forcefully taken away from school, her father cannot bear to look at her futile struggle to escape. He closes his eyes and moves away and tells Mindak that the *gale* (cloth worn on the lower portion of the body) which falls to the ground should be given to her. Though very weak and insignificant in comparision to society and social norms, Kargum is very much affected by what happens to his daughter. He talks to himself:

> She might have thought that I had come to give her money. She came out running with so much hope. How could she think that I had brought a group of people to carry her away forcefully? But it is a social custom. I have given her in marriage to them. And they have the right to take her away (*Kanyar Mulya*: 39).

Kargum is not at all happy after Gumba is dragged away from the school. There is a lot of conflict in his mind and from that day itself he loses his appetite (*Kanyar Mulya*: 48).

Of course, contemporary readers may have doubts about Kargum's love for his daughter as he takes bride-price for Gumba and sells her. However, we can only do justice to Kargum's character when we look at him from the point of view of the traditional Adi society of those days. Kargum often tells himself that he should not have done what he did. He tries to convince himself that he done what he did. He tries to convince himself that he has done the proper thing because after all he has to follow social customs:

> I am angry with myself. Why did I sell her? Every one would have been living peacefully, if I had not sold her. But there was no other option for me other than to follow the social norms. (*Kanyar Mulya*: 59)

Kanyar Mulya is set against a traditional rural setting which lends it a pastoral smell. The words, beliefs and rituals of the Adi society are clearly depicted in the novel from the beginning till the end. Mention of words such as *kebang, dekasang, mosup, dere,* Donyi Polo, Abotani etc. gives the novel the feel of the traditional life of the tribal people of Arunachal Pradesh, especially the Adis. The writer is successful in expressing the traditional ideology very artistically. The situations permit the readers to accept the traditional ideas of Adi society. 'Donyi' and 'Polo' refer to the sun and the moon respectively. The Adis' practice of the worship of the sun and the moon is frequently mentioned in the novel. They are shown to generally have deep faith in various gods and goddesses, spirits and witches. Each and every action, both inside a house or outside it is explained as a wish, power or punishment of some spirits, good or bad. Dai gives such minute details of some of the social customs of Adi society in the novel that he allows the reader to not only have a feel of the society, but also to feel as if he is a member of that society. Mention can be made of a particular social rule of Adi society. According to the traditional custom, in the name-giving ceremony of the Adis the last syllable of the child's father must be the first syllable of his son or daughter's name. For example, Kargum's father's name was Tumkar. The 'kar' of Tumkar was added to 'Gum' and so he was named his son as Kargum. Similarly Kargum's daughter is named Gumba (Gum+ba).

There are not only mentions of the system of bride

price which was prevalent in the Adi society, but the very theme of the novel revolves round this particular custom. Kargum's thoughts clearly reflect this:

> He got five mithuns, seven cows, eleven dishes and six brass bowls as Gumba's price. And he got three thousand rupees in cash. (*Kanyar Mulya*: 2)

There is frequent mention of *apong* (rice beer prepared at home) and its importance in the Adi society. It is customary to offer *apong* to guests in Adi households. When people gather in Mindak's house for the first time to discuss Gumba and Dakat's (Mindak's son) marriage, *apong* is served to each and every one. The preparation of *apong* is another custom that is mentioned (*Kanyar Mulya*: 90). When Kargum goes to Itum's house to ask for the latter's daughter's hand in marriage for his youngest son, he is offered *apong* by Itum and his wife.

The mention of *kebang*, which is the main judicial platform for the discussion of all cases, gives a vivid picture of the Adi society. In the Adi society, grinding of rice with the help of *ural* (traditional instrument for grinding rice) and gathering firewood from the forest are two major works of the women. Such works are also done by Gumba. Here too, Dai is succesful in presenting his heroine against a completely rural setting where a traditional Adi girl like Gumba is expected to be born and brought up. Therefore, we see Gumba helping her sisters-in-law in grinding rice just after the day when she is told that she has been sold to someone. Even though she is mentally very upset, she does extend help to her sisters-in-law in grinding rice. Then she goes to the forest to collect firewood too. The writer portrays a true picture of an Adi girl proceeding toward the forest:

> Gumba took a 'hora' from the upper berth and kept a big knife and a spear inside it. Next she considered which way she should go to collect firewood. She thought for a while and then started walking. (*Kanyar Mulya*: 18).

Another important Adi tradition that is depicted in the novel is the system of dividing rivers and forests between individuals for fishing and hunting. Kargum murmurs about Mindak.

> How many rivers does he have to eat fish from? How many forests does he have? But I do have enough. I'll see, who will people sing for in times to come? (*Kanyar Mulya*: 27).

Another tradition of the Adis is that they cut the ears of mithuns to give some marks when they are young. From the marks of the ears one can recognise its owner. When Kargum is in search of mithuns he observes:

> He could clearly see that a mithun was eating leaves. But he was depressed on seeing its ear-marks. It was not his. (*Kanyar Mulya*: 28).

Hunting is a popular tradition of the Adi society. In a traditional Adi society, hunting for the purpose of eating is a usual practice. Kargum too is quite good in hunting:

> He went out in search of deer and saw something running fast. He lifted his gun but before he could press the trigger, the deer had crossed the hill (*Kanyar Mulya*: 29).

The prevalence of class-system and slavery that was prevalent in the Adi society deserves mention here. Dai mentions these on different occasions in the novel. These were prevalent in the Adi society before the beginning of British rule to keep the orphans or the people who could not repay the debts or the war-prisoners as slave. The readers are introduced to such old customs when Jokom says, "Talk sensibly, you are here to talk just because the government has set you free" (*Kanyar Mulya*: 22). It is mentioned in the foot-note of the novel that even after independence, slavery was prevalent in the Adi society. But in the last decade, government has set all the slaves free. Leeter, a classmate of Gumba, says, "We have to thank our society that the very cruel system of slavery has been abolished ..." (*Kanyar Mulya*: 47). This indicates that it was prevalent in earlier times. Once when Kargum asks Minjum to write to Mindak, he says:

> None from my clan has ever been kept as slave. But many from your clan were kept as slave by others. Even in the houses of my own clan-members, many people of your clan used to work as slave (*Kanyar Mulya*: 48).

Borgam Gaken talks about the system of slavery at the end of the novel:

> Till a few years ago, we had slaves. But now the slaves are living a free life (*Kanyar Mulya*: 99).

Upar Mahal (2003) is the last novel written by Lummer Dai, completed just a week before his death. According to his wife Nanni Dai, the day he completed the novel, he fell sick and died within a week (Preface). *Upar Mahal* is quite different in manner and subject matter from his other novels. In all his previously written novels he deals with the Adis and the Adi social life, their customs and rituals traditions and beliefs. Here, we see a totally different picture of the new generation - a picture of the Arunachalee society in general. The social transformation taking place in the society of Arunachal comprising different tribal people as well as non-tribals is the main focus of the novel. In *Upar Mahal* too there is mention of different customs and traditions, rituals and cultures of different tribes of Arunachal, but the author soars artistic heights to vividly showing the rapid social transformation taking place in the society. The young lot with their new values and the beliefs of a money and power-driven society can be seen in different characters and their dialogues and actions. In almost all his previous novels the author presents people who are simple by nature, untouched and as yet unspoiled by the rigors of modernization and present day existence, thus preserving their rich cultural heritage in pristine form. Things are however changing rapidly.

Though there is mention of Donyi Polo (the Sun and Moon whom they worship as God), *apong, gale* (the cloth worn by the ladies in the lower portion of the body) and *galuk* (jacket worn by man), the Adi *apong*, the Khamti *lungi* (*Upar Mahal*: 1), the Buddhist tradition, butter tea (*Upar Mahal*: 20), *surpi* (a food item like butter), *ekung* and *onger* (*Upar Mahal*: 51) , Solung Festival (*Upar Mahal*: 55) Ponung dance (*Upar Mahal*: 58) *mosup* (*Upar Mahal*: 75) and others, the important issue that the writer is dealing with is the sudden and drastic change taking place in the Arunachalee society. He is concerned with the thirst for money and power in man.

Jalam's thirst for money and power leads him every time to attempt to kill anyone whom he considers to be an obstacle in his way. He threatens to kill everyone and on every occasion, his adversary agrees to do as he says for fear of losing his life. Everytime he tries to kill people from the 'upar mahal' (by which he means the "higher level" including the ministers, chief minister or any top-level bureaucrat of the government departments (*Upar Mahal*: 27) when they stand in his way. On the first occasion Jalam wants to kill Premjit Yadav, the junior engineer whom he requests for a contract and says, "I have to get what I want. I want money to become an MLA and then a minister ... So you have to convince the higher level" (*Upar Mahal*: 12-13). Then he attempts to kill another member of the 'upar mahal': "I actually had a different plan ... to kill you; I am prepared to be hanged for killing you. I am not scared of being hanged"(*Upar Mahal*: 14). Jalam gives him Rs. five lakh as a bribe for the contract. But when the higher level (the official) pretends to forget about the money, Jalam fumes in anger and attempts to kill him. In fact, Jalam keeps threatening to kill his adversaries. Jalam has to face a lot of adversities, both in his political and familial life. And he can not solve his problems by honest means. His threat of killing people is sometimes just a ploy and sometimes a spontaneous outburst of his anger.

Upar Mahal basically deals with the protagonist Jalam and his life, thoughts and activities through which we can trace the growth of a modern upcoming youth of the present society.

Jalam continues to be a womanizer even after his marriage. This affects not only his domestic life but his political career too. His friend Sonam's sister Drema is his beloved with whom he shares many sensuous nights and as a result she gives birth to a child. Next, he develops physical relation with a dumb labourer whom he names Yabin. He gives money to Yabin to take care of her child. After Yabin it is Yaka who sleeps with him for a job and conceives. She dies at the time of delivery. But till the end of the novel there is no mention of Jalam's repentance, instead he proclaims,

> Even if people call me a mean murderer or a corrupt man, I am not the one to bow down before others ... I belong to a powerful clan. I may be a wicked man but the prestige of my family name will help to keep my image intact" (*Upar Mahal*: 215-216).

But at the end of the novel he is weighed down by various critical problems - both at home and as a home-minister. The realization of the mental suffering of his wife ultimately leads him to try to kill himself with a pistol. Being involved in one crime after another, Jalam transfers Inspector Wangkam to make way for OC Tapam as the new Inspector to kill Yaka's brother-in-law who is in police lock-up.

Upar Mahal is quite different from the other novels by Lummer Dai - particularly in its theme and content. It can be considered a political novel though it is not affiliated to any particular ideology as its main focus is on the political life of a person as well as the hypocrisy, inhumanity and brutality associated with today's politics.

In the course of his career there is a remarkable change in the writings of Lummer Dai. *Upar Mahal* was published twenty four years after the publication of *Kanyar Mulya*. Unlike Dai's other novels, where the characters are mostly from the Adi tribe, the characters of *Upar Mahal* are from different communities, including both tribals and non-tribals. The society of *Upar Mahal* is a newly formed multi-ethnic society. He writes about an emerging class of people in white collar jobs whom he identifies as belonging to the *upar mahal*. With this novel Dai also makes a remarkable departure from the traditional rural setting to the world of sleaze and decadence in a modern day town such as Itanagar.

Section B

Yeshe Dorjee Thongchi, another prominent figure in the field of Assamese literature, is an Arunachalee. He has written four novels and three short-story collections till date. His novel *Mouno Ounth Mukhar Hriday* has won the Sahitya Akademi award in 2005. His novels are:

Sonam (1981)
Lingjhik (1983)
Mouno Ounth Mukhar Hriday [Silent Lips Animated Heart] (2001)
Saba Kata Manuh [The Man Who Slices Dead Bodies] (2004)

Yeshe Dorjee Thongchi's narratives paint brilliant landscapes of Arunachal Pradesh - the land of dawn-lit mountains, dense forests and rivers - unfolding a panorama of scenic beauty with variety of people and their cultures and traditions. Most of his novels deal with the traditional, cultural life of the Monpa and the Sherdukpen tribes. The appeal of his novels lies in its being rooted in an indigenous tribe and its culture and at the same time bearing a pervading note of universality.

His first published novel *Sonam* centres round the Brokpa tribe (a section of the Monpas). The family system of the Brokpas where a woman is socially allowed to marry more than one man is the custom in which the story of the novel is rooted. In such societies, polygamy and polyandry are socially permissible and the trauma, agony and the conflict of the characters can be traced to such system due to which the man-woman relationship is affected.

The cultural setting of the novel (i.e. in the community at the backdrop of Tawang and Dirang of Arunachal Pradesh) demands the mention of the life-style, culture, and customs of the society where the domestic yak-oriented economy is the mainstay of the locals. *Sonam* begins with the describtion of a robust young shepherd of the particular community coming back home from the Choksum hills to the Sakteng village. His name is Lobjang. To portray the Brokpa life and their details, the writer introduces us to many tribal words and traditional customs of the tribal people of the region, especially the Brokpas. It has been already mentioned that the main theme of the novel centres round the social system where a woman is married to more than one man. Apparently it seems that women enjoy more freedom, but while going through the novel it becomes clear that the community is really male-dominated.

Polyandry and polygamy are permissible in Lobjang's community and we hear him talking to his wife to get another groom for her in a matter-of-fact way. When he meets Pema Wangchu with his wife at his house, he says, "If your mind is set on Pema Wangchu let us bring him into our household" (*Sonam*: 9). Again he says, "If you do not want him to come and stay in our household, we can search for another groom" (*Sonam*: 9).

It is clear that Lobjang is not at all worried or disturbed to share his wife with someone else. Polyandry is prevalent in the society of the Brokpas to which he belongs, and he is so constructed by the ideologies and beliefs of his society that even the idea of sharing his most cherished woman does not create any trauma or agony in his mind. Instead, he is happy to get a man who would help him in his household activities.

Moreover, the mention of *chupa* (people or the

panchayat), *aara* (liquor), *gorbu* (vessel), (*Sonam*: 14); *nasang* (permanent host and friend, *nekar* (pilgrimage) (*Sonam*: 26); *duntsang* (to celebrate by dancing and drinking), *saksoiban* (to bow down) (*Sonam*: 28); Rimpoche (the abbot appointed in Tawang monastery by His Holiness the Dalai Lama), *khata* (a piece of cloth traditionally used in ceremonies) (*Sonam*: 29); *drepa* and *anei* (hermit and nun) (*Sonam*: 31); and *boitang* (Tibetan coin) (*Sonam*: 84) and so on, proves the author's close attachment to the society about which he is writing.

The novel reveals the influence of Buddhism on the writer. A religious note pervades the entire novel. We sense the deep faith of Lobjang in religion when he takes his children Rinchin and Tasi to the Tawang *gompa* (Buddhist Monastry) to make them *anei* and *drapa*. Lobjang and his family's pilgrimage to the Tawang *gompa*, their performance of different rituals such as roaming seven times round the *gompa* taking maize seeds from *changje*, who is the secretary to the abbot (Sarkar, 1980:41) in a *khata* and to throw it on the statue of Sun-Raji (Tara Devi) (*Sonam*: 29) - tell us about Buddhism and the rituals associated with it.

We have an elaborate description of hunting as an important occupation or a favourite passtime of the tribals in the novel (*Sonam*: 69). The belief in rebirth among the tribes is also reflected in the birth of Rinchin who used to speak about her former life as the pet dog Dombu.

The author is successful in drawing the picture of a pastoral society while describing the incidents in terms of some traditional beliefs, superstitions and cultural system. When the tiger which has killed many animals of Lobjang and others cannot be killed, the people of the village are frightened and start to believe that it is not the doing of a tiger but some spirits who have come to take revenge on account of some misdeeds of humans. Finally the tiger is killed by Lobjang and Tikarai. When they come back to the village carrying the corpse of the tiger, both of them are given a traditional welcome.

Thongchi talks about the different ways of worship of traditional deities; for example, when Sonam dies her mother puts a *boitang* in her mouth and a *khata* on her body.

Patriarchy is quite prevalent in the society which is reflected in the characters' voices in different situations. At the very outset, thinking of an age old belief of violating the sanctity of the farm because of the presence of a woman coming to the farm and staying together with her husband, Lobjang rejects the idea of bringing Sonam to the farm. There is no doubt that this expresses the strong note of patriarchy prevalent in the society. We hear the voice of a male chauvinist when Lobjang tells himself on the way to Pema Wangchu's house:

> Women are like dogs; if anyone pats a woman on her back - she will meekly follow him, and will snarl and snap at him who once in his life speaks or behaves roughly with her (*Sonam*: 62).

On another occasion, his male ego does not permit him to go to Pema Wangchu's house and have dinner there in spite of the invitation of his wife. He refuses Pema Wangchu's sister and had dinner in Tikoroi's house. (*Sonam*: 75).

However, in spite of all these there is relationship of love and care between Lobjang and Sonam. Each of their solliloquies is proof of their great love for each other (*Sonam*: 80). Sonam breathes her last on Lobjang's lap and the theme of love is highlighted once again when she declares: "Today I can say that my name 'Sonam' has not gone in vain" (*Sonam*: 83) ('Sonam' means 'the fortunate one').

Lingjik (1983), the second novel by Yeshe Dorjee Thongchi, is basically about an old custom which was almost dead by the time in which the novel is set. But this old custom is given a fresh lease of life as a result of a clash between two persons leading to a clash between two clans. 'Lingjik' means 'a pillar of stone'. The

Sherdukpens put the 'Lingjik' in the ground as the symbol of a very important decision or as the line of boundary to remind the heirs of the clan about that particular issue. They believe that if anyone violates that decision, the stone pillar would harm him/her. *Lingjik* is a novel where the author has talked about many rituals, customs and traditional beliefs which are not just commentaries on Arunachal's tribal life but also help to carry forward the story. This novel, though purely local in colour, comprises a variety of themes which makes it scale artistic heights. There is no doubt that the appeal of the novel lies in it being rooted in the indigenous Sherdukpen custom and culture but it is also true that it has an all-pervading tone of universality due to its different themes such as the social change/transformation taking place in the society; conflict - both inner and outer; the process of identity formation in society by which the human mind is framed.

Tribal societies are usually believed to be full of beliefs coming down from time immemorial. Thongchi is able to present these beliefs in the course of the progression of the story in a casual, almost matter-of-fact manner, for instance:

> There is a proverb that the mother whose first child is a girl is always lucky (*Lingjik*: 9).

In the discussion between Penje and Darja there is a string of references to such beliefs and superstitions. In answer to Darja's question Penje says, - "There were many reasons for offering *puja* in their household. There is bound to be diseases in a household where *pujas* are not held" (*Lingjik*: 60). At one point in the novel Lamdoima happens to faint and immediately after that a priest sits beside her and proposes to offer a *sur* (worship of spirit to prevent it from harming human beings) which represents the society's faith in such beliefs (*Lingjik*: 105). Towards the end of the novel, when Krima comes home, thinking of her brother Premtasi's marriage (though he has invited her to see their ill mother), someone keeps some pieces of burning firewood near her feet and she thinks, "Why was the pieces of burning firewood kept near her feet? Was there actually any sick person at home?" (*Lingjik*: 155).

The author is quite capable of mingling art with life which prevents the novel from being a monotonous social history of a particular community or just a description of different incidents strung together. At the very outset of the novel the readers are introduced to the traditional Sherdukpen marriage system through the words of Premtasi:

> Suddenly some people from the groom's family drop in and take the girl to the groom's house ... After keeping the bride at the groom's family for three days, she is sent back and again on an auspicious day she is brought back by the whole of the groom's family (*Lingjik*: 4).

Premtasi also informs us about another aspect related to the marriage systems of the Sherdukpens when he says, "There can be marriage between maternal cousins in our society" (*Lingjik*: 5). In the Sherdukpen society if a niece gets married to her maternal uncle's son she is called *samu* (*Lingjik*: 16). People prefer such "readymade brides" (*Lingjik*: 16) as they are known to the family. Krima is one such *samu* of Penje who is her maternal uncle and therefore he has complete right over her (*Lingjik*: 17).

There is mention of a particular bird called Bobo Thekong Kong and the writer has explained (in the footnote) the folktale associated with it. According to this folktale the Bobo Thekong Kongs were very ugly but the crows were beautiful and fair in complexion as well as very good painters. One day Bobo Thekong Kong said to the Crow, "Dear friend, please draw some beautiful pictures on my body and I'll do the same for you". The crow did as asked and expected Bobo Thekong Kong to paint him beautifully. But the latter painted the beautiful crow with soot. Being very angry the crow broke a leg of the Bobo Thekong. Since that day Bobo Thekongs are scared to leave their nests during the day time and have become nocturnal. When Prematasi goes to his sister's house at 2 p.m. after the discussion with his uncle and aunt - Penje and Ajom about Krima - he compares himself to Bobo Thekong Kong - someone who had betrayed those who loved him (*Lingjik*: 38). With the help of appropriate references, the author introduces the readers to such folktales very tactfully.

The subject matter of the novel demands restoration of a particular social custom of the Sherdukpen community - the custom of putting a stone pillar called 'lingjik' in the ground as a reminder of a momentous decision or as a line of demarcation between two persons just to remind the next generation about that particular decision. Along with this, there is renovation of another old custom i.e. to take bride price during the life time of the woman. Significantly it is considered to be the most insulting issue for the woman. In the present novel, Krima is brought back home by Premtasi, her brother blaming Dorjee Sewang for his impotence. This not only hurts but also insults Dorjee's family and all the people belonging to the Wangja lineage come to ask for the bride price from Kejang for Lamdoima. Lamdoima, Kejang's wife is unable to bear such insult causing her acute mental anxiety and this is what ultimately leads to her death. The system of taking bride price in the life time of the victim is believed to be very shameful and insulting. Many other customs are related to the one of asking for a woman's bride price during her lifetime. For instance, there is the custom of using a middle man called *khalapu* (*Lingjik:* 123) to carry forward the process of asking for the bride price; the party that asks for the bride price settle down for a few days in a place near the village of the other party and tries to ascertain the price of the bride in terms of brass utensils, agricultural land, horses, one bullock or two cows, five bowls, five pieces of clothes of the endi material and a little bit of silver jewellery (*Lingjik:*125).

There is mention of a game called "holding hand". The author describes this game in the footnote as a very popular game of the Monpa and the Sherdukpen tribes. In this game, one grasps the other's hand while the later tries to be free from the former without taking help of the other hand. It is a game to test the strength among the young men (*Lingjik:* 133). In the novel, a man from the Wangchu lineage offers a hand to a member of the Thonggu lineage after the quarrel between the two parties. At the same time, the pledge by the Wangja and Thonggu parties beside the *Lingjik* reflects their faith in traditional social customs and the influence of such customs over the common people:

> We touch the *lingjik* and take a pledge today that from now on we have snapped all relation between the lineages of the Thonggus and Wangjas and if in the future any person dares to defy this, then the *lingjik* will harm him/her (*Lingjik:* 138).

In spite of the highly traditional feel of *Lingjik,* one also finds a sense of urge for change in the society. The social transformation taking place in the society is also reflected in various incidents at different places. Premtasi, an extra Assistant Commissioner in a town is the embodiment of such a transition.

The very beginning of the novel provides ample evidence of transition. Premtasi and his affluent friends assemble in someone's house and drink whisky and utter 'Cheers!' which is a completely new scene in the traditional tribal society. At the end of the first chapter, when Premtasi decides to break the bond of marriage between his sister Krima and Dorjee Tsewang which was held according to the traditional way of abducting the girl on the way, he strongly protests against this established social rule (*Lingjik:* 7). Here we sense the change or transition coming about in society.

Premtasi's dream is to educate his sister Krima and so he tells his uncle and aunt (Krima's in-laws),

> Our tribal women will never be able to enjoy the benefits of higher education till the day we continue with the system of abducting girls from the road to make them brides (*Lingjik:* 18).

He decides to 'keep pace' with the modern society as he believes that it is necessary to get rid of the old customs and superstitions for the betterment of the society. At the same time, he believes that he must begin by changing things at home (*Lingjik:* 21).

Premtasi wants to change the marriage system of his community (*Lingjik:* 41) - it is obviously because of the transition in the society that someone has the courage to speak against old customs. Premtasi appoints a tutor for Krima (a new step to educate a girl) and buys a record player and radio to compensate for her loneliness. These are modern means of entertainment which definitely help to convey to us the fact that the tribal society is going through a period of transition.

Norbu Tsering is another character who fights in favour of change in society. He and his group meet Penje and talk a lot about the change that should be brought about in the society (*Lingjik:* 18). They request Penje to do away with the old custom of taking bride price during the life time of a woman (*Lingjik:* 88). A great change is observed in Krima's words when she loses her mother. She scolds Uncle Penje and when Premtasi asks her as to why she did not feel ashamed to do that, she says:

> I do not have any sense of shame. Can a wooden doll have any sense of shame? I am a wooden doll without a mind of my own. Therefore, you, my father, my uncle everyone is playing with my life - without being least bordered about me, my decision, my mind. You males are just like beasts - you do not consider women as human beings (*Lingjik:* 160).

Such words by Krima has a tremendous impact on

Premtasi's mind due to which he accepts his guilt of going against so many people and hurting others. What is more significant is not only the change of outlook in Premtasi, but the courage and strength that Krima has shown in talking so frankly about such issues. This must be considered to be a sign of the changes setting in in society.

In the Sahitya Academi award winning novel *Mouna Ounth Mukhar Hriday* (2001), Yeshe Dorjee Thongchi adopts the universal theme of love with an uncoventional treatment. The story is set in the North East Frontier of Assam (NEFA) of the 1950s. During the construction of a new road in one of the hilly areas of NEFA, two young people - a boy and a girl - belonging to different tribes fall for each other. The novel deals with the human feelings of love, hope, aspiration, and dreams which transcend the limit of society and even that of language between two lovers. Both Rinchin Norbu and Yama Natung are engaged to someone else according to their own customs and traditions. Even then, the intense love between the two continues to blossom. But at the end of the novel, they can not overcome the social barriers to be happily united in wedlock.

We are introduced to Rinchin's love for Yama at the end of the second chapter when we find him trying to spend his first night at the site of road-construction. He had met Yama during the day in the forest where they had gone to collect firewood:

> The incident in the twilight, the embrace of the warm and soft body running to him in fear, the two tearful eyes, the act of gently filling up the *hora* with kindling, the sensuous gait of the body walking ahead of him, the soft and shapely hand pointing toward the camp, the sweet smile intended for him at the departing moment - all these images flashed one after the other in Rinchin Norbu's mind and left his body and mind in a state of pleasurable excitement. (*Silent Lips Murmuring Hearts*: 19-20).

On the other hand, Yama also experiences similiar emotions which are completely new to her:

> Those eyes .... that face ...eyes like that, where had I seen eyes like that? I must have definitely seen them somewhere. There can be no other reason for them to seem so familiar and so personal. Love ...maybe this is love. Is it that I have fallen in love with that stranger? (*Silent Lips Murmuring Hearts*: 22-23)

Rinchin is fully aware of the fact that he has to marry Tsering Jangmu and he decides to forget Yama, a girl from another community. However, he can not help loving Yama and meeting her.

There is a series of incidents (and coincidences) that takes place helping to bring the lovers closer. At first when Yama is about to be killed by Tsering Jangmu, Rinchin saves her. Again during the time of road construction, when Rinchin is about to fall from the hill it is Yama and her presence of mind that saves his life. It is for the sake of a bunch of *lesang mintu* flower that he wants to give to Yama that Rinchin puts his life in jeopardy (*Silent Lips Murmuring Hearts*: 57).

The title of the novel expresses the deep love between two souls where speech is insignificant. Both the lovers speak different languages unknown to each other but that does not seem to be a barrier in their communication. The lovers converse silently with each other. Yama has immense faith in Rinchin's love:

> I know you love me. I can see that in your eyes and in your look (*Silent Lips Murmuring Hearts*: 78).

The same idea is expressed by Rinchin: "In your eyes and your thoughts I can discern your love" (*Silent Lips Murmuring Hearts*: 79).

On several occasions the lovers try to forget each other considering the difficulties associated in the way of their love, but neither of them can manage to do so. Therefore when Tadak asks Rinchin to forget Yama, Rinchin really thinks of obeying him: "He wanted to give his word to Tadak but a part of his mind rebelled" (*Silent Lips Murmuring Hearts*: 83). Similarly Yama says: "I have decided many a time that I am not going to think of Rinchin anymore; but the more I try to forget him, the more I tend to think of him" (*Silent Lips Murmuring Hearts*: 136).

Though Thongchi deals with the universal theme of love in this novel, he does not fail to present before us a wonderful picture of tribal life and society. The novel is full of folk beliefs, superstitions and cultures and we hear about these either in a character's dialogue or in the form of narration by the author himself. In the very first page of the novel the Sherdukpen girls express fear of *bonmanuh* and *jokhini* (spirits or ghosts). Even the bold and daring Rinchin has shadows of suspicion lurking in his mind when he encounters Yama for the first time:

> I have never seen anyone as beautiful as her in my life. It was beyond my wildest fancy that any human being can be as good-looking as this. Is it *bonmanuh* or some *jokhini* of the jungle in the guise of a girl? (*Silent Lips Murmuring Hearts*: 3)

One of the Sherdukpen girls, Wangmu, narrates a "true story" about *bonmanuh*. All the Sherdukpen girls firmly believe that Yama is a *bonmanuh* and so they surround Rinchin on the way because they think that "in the presence of females, the female *bonmanuh* becomes harmless" (*Silent Lips Murmuring Hearts*: 6).

As Rinchin and Yama are from different tribes i.e. Sherdukpen and Nyishi respectively, the novel has an account of the beliefs and customs of both the tribes. The Sherdukpens used to go to Assam in winter to a place called Doimara where they made friends with whom they exchanged commodities. (*Silent Lips Murmuring Hearts*: 14). The marriage system of the Sherdukpens is mentioned in different situations of the novel. Tsering Jangmu is engaged to be married to Rinchin. We are told at the very beginning that the initial rituals of marriage had been already performed. Tsering Jangmu was forcefully caught and kept in Rinchin's house for three days according to the custom. According to the custom, she was not supposed to eat anything in Rinchin's house during these three days and he had to be away from home for this period (*Silent Lips Murmuring Hearts*: 40). It was now just a matter of Rinchin's family formally welcoming her to their house and the marriage ritual would be over.

Thongchi also talks about the Nyishi marriage system of giving bride price in the form of mithun. When Rinchin asks Tadak about his marriage he answers:

> I have already given the bride-price – five mithuns, and we have to give more. (*Silent Lips Murmuring Hearts*: 47).

Even in Yama's words we hear about this system,

I am a commodity that has already been sold. I have no identity of my own. Nor do I have a voice or view of my own (*Silent Lips Murmuring Hearts*: 78).

The traditional way of welcoming guests in the tribal societies presented in the novel is by performing songs and dances and offering them *laopani* (local wine). This is done when the engineer Lal sahib comes to check the progress of work.

During the Governor's visit, the tribal youths present a variety of traditional dances. The Monpa and the Sherdukpen dances attract the audience's attention with their unique styles. The Sherdukpen's eating habit, including their resistance to the mutton of domestic goats, is mentioned in the big feast organized among the labourers of all tribes.

In the NEFA of the 1950s communication was obviously quite difficult and challenging as there were no roads and people had to cross steep hills on the way. There were porters booked by the government to carry V.I.Ps and their luggage. In the traditional way of life, the tribals were naturally accustomed to hunting and fishing. The author has introduced us to such customs through Tadak and Rinchin's characters. Tadak always carries bow and arrows with him. On the way back home from Bomdila, both Tadak and Rinchin go hunting and fishing (*Silent Lips Murmuring Hearts*: 158-159).

The author has tried to add to the realistic picture of the variety of tribes and tribal societies of Arunachal with the inclusion of the different words from this world such as *laopani, apo, harang, ynichak, gidi, duchu, nida, chape and chanap, mone, tulu, yna, Libing* etc.

Arunachal, the land of rising sun, is famous for its scenic beauty. The beauty of the virgin hills of Arunachal Pradesh finds a mention in the novel. In fact the beauty of the landscape supplements the feelings of love that has crept into the hearts of Rinchin and Yama. The author describes the beauty of nature in Rinchin's words:

> After walking some distance he saw a high mound. In the centre of the mound was a pine tree. When he stood beneath the tree he observed the darkness of evening gradually enveloping the world. Light and shade were playing hide and seek through the branches and foliage of the pine tree. There were a few unthreatening clouds scattered in the sky and they had the look of cotton that was sparingly spread by somebody. The crimson rays of the evening sun fell over the clouds giving them a ruddy hue and the panorama was supplemented by the uniquely serene sky of autumn (*Silent Lips Murmuring Hearts*: 43).

It has been already mentioned that the novel has a realistic feel which adds to its appeal. The readers are given the idea of the time and the place in which the novel has been set. The narrator describes:

> It was six or seven years ago in 1951 or 1952 that the thick forest of the Bomthe hills were cleared to develop Bomdila as the headquarter of the

> Kameng Frontier Division and the offices were shifted from Chariduar to this new township. (*Silent Lips Murmuring Hearts*: 144)

So, the story is set in the NEFA (Noth East Frontier Agency) of 1957-58 when the place was almost cut off from the rest of the world and it seems that the author succeeds in evoking the time of that period. The construction of the road by volunteers from Arunachal Pradesh by making their way through the high hills of the state is the immediate reason for which Rinchin and Yama come to the Saku hills. The characters of the novel who are tribal people from the same state, actually speak different languages. They have different rituals, traditions and customs. Though living in the same state, one's language, culture and customs, dresses and food habits are totally alien to the others.

The description of Rinchin's first experience in a motor vehicle and his description of Samtan's (an old man) encounter with a motor vehicle give a realistic picture of the innocent tribal people of the fifties. Rinchin describes Samtan's encounter with a motor vehicle in the following words:

> If you were to ask Chamtan, he will tell you that the ghost [the motor-car] came running after him, devouring tree leaves on the way (*Silent Lips Murmuring Hearts*: 17).

This gives us a fair idea of the people of Arunachal Pradesh. The writer gives a realistic description of the people's first experience with rupees:

> People like Takar, Dorjee, Chouky, Yama, Yado, Tsering Jangmu who chanced to see currencies for the first time in their life, placed them on their hands to smell the mild scent being emitted by the notes. They minutely observed the images of the Ashok *charka*, Mahatma Gandhi and the letters printed on the notes. They marveled at the fact

> that exactly the same images were there on all the notes. (*Silent Lips Murmuring Hearts*: 89)

Their first experience in a shop, after they come to know that they can get things in exchange of the money that they had been given by the *mohori babu* for their work, is very interesting. Though quite amusing these were the kind of experiences that the tribal people of Arunachal Pradesh, who were totally cut off from the outer world, underwent in reality. Yama's astonishment after looking at her own reflection in the mirror is very amusing but it is also realistic.

Though NEFA was a land inhabited by different tribes, many of these tribes were totally unknown to the others due to geographical isolation and difficulties in communication. The language, rituals, traditions and customs, dresses and food-habits of one tribe are different from the others. The author has tried to give a realistic detail of the same in his novel.

The Nyishis believe themselves to be the most superior people in the world (*Mouna Ounth Mukhar Hriday*: 23-24). We can hear the same kind of feeling in Rinchin's words when he is talking to Tadak:

> Even though we all have eyes, mouths and hands we are all different from one another. We Bhutias think that we are the greatest race while you the Nyishis think that you are the greatest. That is why we hate each other. We consider each other as enemies and fight (*Silent Lips Murmuring Hearts*: 123).

The writer looks at another issue very minutely – that is the existence of true human values, simplicity, and a feeling of love and camaraderie between Tadak and Rinchin's societies. Tadak and Rinchin's friendship speaks volumes about the degree of honesty of which they are capable.

Moreover, Tadak and Rinchin's conversations in which they share their sorrows and happiness with each other, in spite of their difference of views, are very significant. Rinchin confesses his love for Yama to Tadak. Tadak, instead of being angry with Rinchin (which was quite natural in such a case), listened to him sympathetically. Tadak advised Rinchin to forget Yama considering the social barriers that separated them (*Mouna Ounth Mukhar Hriday*: 118). Even after coming to know about Rinchin's feelings for Yama, Tadak invites him for dinner (prepared by Yama) to his camp and accepts the latter's request for extending their stay till the Governor's visit.

Tadak and Rinchin together drink the wine that has been given to the latter by a sepoy in exchange of a piece of venision. Tadak requests Dilip Saikia to write an application to the Political Officer for Rinchin, asking for a license for gun. One can see that at the end of the novel, in spite of the difficult situation in which they are placed that does not allow them to make a public display of their emotions, they part from one another with heavy hearts.

Set against the background of the Monpa society of the 1950s and with references to different social, historical and political happenings of the time, Yeshe Dorjee Thongchi's *Saba Kaka Manuh* is about a man whose profession is to slice dead bodies into pieces and his love for a saint. Thongchi has said something very interesting about this novel:

> It is a rare experience for me to write this novel. After writing this novel, I feel that I am unable to complete the other novels I am writing. Actually, *Saba Kata Manuh* represents the peak of my career as a writer and anything that I write from now on will only be of lesser merit than this (*Satsori*, 2006: 10).

In his own writings about his narratives as well as in the preface to this novel, Thongchi has said a lot about what motivated him to write this novel and about the characters and setting. This gives us enough scope to discuss the thematic concerns of the novel. The most important theme of the novel (as per his admission in the preface), is the system of cremation of the dead by slicing the body into a hundred and eight pieces and then to throw these pieces in to the river, that is prevalent amongst the Monpas. The incidents and the story of the novel revolve around this custom. There is an absolutely vivid and realistic description of the man who is involved in the process of slicing dead bodies and of the very process itself. The writer is so realistic in his description of the *thampa* and his work that the readers are actually made to see these with their eyes. While going through the novel one can feel the mental state of the *thampa,* one can smell him, and can also hate him for all the "dirt" in his hands - smeared with blood and puss of the dead. Moreover, the realistic details of these incidents make us believe that the realistic representation is another important aspect of the novel. Even the characters of Param Pawan Dalai Lama, the additional political officer T.K. Murthy, and the commander of the army, Lieutenant General Niranjan Prasad, are all based on real people that Thongchi had met in real life as confessed in the preface.

The love between the *thampa* (the man involved in the task of slicing dead bodies) and the saint is another important thematic concern of the novel. The love between these two characters is not the usual kind of sentimental love affair between two youths but something much beyond and also very different from that. There are references to a variety of traditions, rituals and beliefs of the Monpas. The significant role of Buddhism in the Monpa society and the immense impact of the Chinese aggression are important matters to be discussed.

The first sentence of the novel sets the tone for one of the themes:

> The actual name of the man is Darge Norbu (*Saba Kata Manuh*: 7).

The novel is about Darge Norbu who is presently known not by his name but as Ao Thampa alias Apa

Thampa alias Ajin Thampa. 'Ao' stands for 'brother', 'Apa' for 'father', 'grandfather, old man etc. 'Ajin' means maternal uncle and 'Thampa' is a man who cuts dead bodies (*Saba Kata Manuh*: 7). So he is a man who slices dead bodies into a hundred and eight pieces according to the Monpa custom. Darge Norbu has lost the identity associated with his name and has become Ao Thampa for everyone.

It has already been mentioned that the story of *Saba Kata Manuh* revolves around the typical Monpa custom of slicing dead bodies into a hundred and eight pieces. From the beginning to the end of the novel, there are descriptions relating to this particular custom:

> With the first blow of the knife the lower portion of the wrinkle is separated ... then his knife proceeds to work on the other parts of the body." (*Saba Kata Manuh*: 16).

While describing the act of dismembering the corpse, the writer shows tremendous skill in realistically painting the horror involved in the process. At the same time the mental instability of the *thampa* and all the emotions related to it seem to be quite natural. Ao Thampa has become accustomed to his profession. He is no more affected by this. But his heart sinks when he sees the dead bodies of beautiful ladies and the foetus in their wombs:

> Their attractive faces, simple and docile eyes, black hair, well - formed breasts ...all these make him feel sorry for them. He feels like giving life to each of the parts of the bodies ... when he sees the lifeless foetus comes out of the body, it pains him a lot. A sense of sorrow and pain completely engulfs his mind (*Saba Kata Manuh*: 17).

His first experience of helping his uncle in cutting the dead body of Meme Chorgen Karma Thingle is an unforgettable day for him. The description of slicing the dead body into a hundred and eighty pieces evokes a lifelike picture in front of us:

> At first the two legs were cut and then floated in the river. This was followed by the hands. The body without the hands and legs and the head looked like a sack full of flour. What a dreadful sight! There was no blood in it as it was kept for three days and three nights after death. Instead there was a foul smelling puss of a reddish-black colour ... The foul smell of the rotten flesh filled the place" (*Saba Kata Manuh*: 84).

He throws up that day as he sees his hands, feet and his whole body soaked with puss and pieces of rotten meat. The same man later becomes practical enough to be able to slice dead bodies with full attention in a matter-of-fact way. Ao Thampa can later on even guess the reason of someone's death like an experienced physician.

Dorjee Thongchi has confessed that he has used particular incidents from real life. There are descriptions of the death of Monpas in the earthquake of 1950, the transformation of administrative power of Tawang from Tibetan government to Indian government in 1952, the Chinese aggression and Dalai Lama's entering India through the district of Tawang, the *kalachakra* puja in Dirang by the Dalai Lama, the death of Monpas carrying luggage to the frontier during the war etc. (*Saba Kata Manuh*: 6). These incidents give a realistic picture of the time and place about which the writer has written. The division of Tawang Monpa into Chowkcham, Dyakpanang, and Pancheng and its subdivisions is another realistic issue about which there are descriptions in the novel.

The Monpas' custom of cutting dead bodies into pieces is followed in the place called Hrormang in the Surthi village. The Chinese aggression, the soldiers and the fear they generated in the mind of common people are all real. The novelist realistically depicts the story against the backdrop of such a situation:

> The *jawans* who had come to fight are not humans - they are just cruel hunters. One can see anger,

> anxiety and lewdness in their eyes ... (*Saba Kata Manuh*: 90).

Besides these, the Chinese aggression and the exodus of people following it can be called a turning point for Darge Norbu as he is forced to seek shelter in another place by the devastating effect of war:

> The deadbodies of the personel of the Indian army were scattered here and there. It was such a horrible sight ... with the deadbodies, the guns, the corpses of the horses, the motor cars that were broken into pieces ... a pile of destruction ... he had seen a few deadbodies in a certain place, all these were either torn into pieces ... (*Saba Kata Manuh*: 124).

He is taken to Tawang by the Chinese soldiers and is engaged in the road-construction work which was going on from Bumbla to Tawang. When he comes back home, there is no one. After a few days his parents come back home with some of the villagers. Many of the villagers die during the war. In fact, it is the war that separates him from Rijomba. Towards the end of the novel, Rijomba's brother Dondu describes the war to his daughter. Here we are introduced to the places Guwahati, Mirza, Diphu and Udalguri of Assam which allows the reader to trace the realistic note in the novel.

The religious note is the most important concern of the novel *Saba Kata Manuh*. The Monpas are Buddhists by religion. As such, while the setting of the novel is based on a traditional custom of the Monpa tribe, it automatically demands reference to rituals and beliefs of the religion followed by them. The Tawang *gompa*, the second biggest Buddhist monastery in the world, the Lamas, the Dalai Lama, Rimpoche are terms and ideas associated with Buddhism and are used frequently in the novel. We are introduced to the religious rituals of the Monpas when we are informed about the pilgrimage that Darge and Rijomba used to go on:

> For Darge and Rijomba, the most pleasurable time is to go on pilgrimage to the Tawang gompa or to Khinmay, Urgaling, Changbu, Thungmen, Brahmaduchung *gompa* etc. (*Saba Kata Manuh*: 24).

The mention of *nekor* (pilgrimage), *wang* (blessings), *kurim* (worship), the *torgya* festival etc. talk about the religious belief of the Buddhists.

The Monpas' belief in Dalai Lama, His reincarnation and His words as the ultimate truth is the reflection of their unflinching faith in Buddhism:

> *Gelukpa, kringmapa* and *karmapa* are different sects of Mahayana Buddhists and the people belonging to these sects believe that to be born as a human being is the result of the good deeds in many births and that to be blessed with a sight of Dalai Lama is the result of the great devotion (*Saba Kata Manuh*: 65-66).

When they are blessed with the visit of the Dalai Lama, the people including Darge and Rijumba vow down and many of them smear their heads with the soil on which He had placed His feet.

The mantras written in the Tibetan language are beyond their understanding; still they are very sacred for them. All that is to be done can be easily materialised by reading such mantras. Moreover, it can heal both the physical and spiritual wounds; it can destroy the power of evil spirit and can wash off all the sins of human beings leading to the salvation of the soul (*Saba Kata Manuh*: 70).

The rituals such as taking *wang* from the Dalai Lama, trying luck with the Lamas and Rimpoches, and the prophecy that the Rijomba does not have the fortune of settling in wedlock - all these are just the part of religious faith. The realistic detail of the incident of *kalachakra puja* by the Dalai Lama is another instance of traditional Buddhist rituals. On the other hand, Ao Thampa the protagonist, after his transformation to a so-called "normal", religious, social man builds a *chocham* (a room for worship) and starts lighting earthen lamps with *ghee* and incense sticks. He makes *changbom* (*Saba Kata Manuh*: 133) with stone and earth in front of his house. He, as well as the people of his village makes a *mone* (a shrine in the shape of a narrow wall) or *chorten* (a stupa shaped structure), (*Saba Kata Manuh*: 134) in Thampanang and the foundation stone for it is laid by Noraljam Ane Sangge, the great monk Rimpoche. Looking after the orphan or any physically disabled children is a part of Buddhist custom. We hear an unmistakable religious note in the words of Ane Sangge when she requests Ao Thampa and his wife to give their physically disabled daughter to her:

I want to take your daughter with me to look after her. I will pray for her well-being, provide her medical attention and educate her to make her an Ane. I will take care of her after your death so that she would not be a woman without a shelter (*Saba Kata Manuh*: 155).

The description of the puja after the completion of *mone chorten* is another instance of the religious element in the novel.

The theme of love figures in the novel in the form of the relationship between Ao Thampa and the monk Rimpoche. Apparently there seems to be no references to love but when one reads between the lines one discovers a deep sense of love between Ao Thampa and Rijomba, the Rimpoche.

Darge Norbu and Rijomba are engaged to be married by their families in their childhood. Gradually there develops an intimate bond between them. The desire of love is further inflamed by the Chinese aggression in Arunachal. At a crucial juncture the two of them get separated from one another. Darge Norbu becomes a *thampa*, and Rijomba, a great Rimpoche, Noraljom Ane Sangge. It is a mere co-incidence that both of them meet towards the end of their life. The novelist is successful in drawing the picture of their love as something almost divine. The deep note of love is felt all through the novel time and again.

When Dalai Lama blesses Rijomba by placing his hand over her head and asks her name, everyone present there is surprised. But Darge is scared:

> For the first time in his life Darge was scared of loosing Rijomba ... Darge did not want to let Rijomba stay away from his sight for even a moment (*Saba Kata Manuh*: 68).

One day before Darge goes to Lumpo carrying the luggage of the soldiers of the signal regiment (*Saba Kata Manuh*: 99), he and Rijomba meet in the dense forest and have a talk. The author describes the beauty of the latter in Darge's words:

> How beautiful is the mole in her nose! It appears to be even more beautiful now that she is crying (*Saba Kata Manuh*: 98)

Darge's description of her beauty and his remark - "You have to get married and set up a house with me, be the mother of my children" and his promise to present her a nice *todung* bought from the *torgya* fair (*Saba Kata Manuh*: 99) are all expressions of his love for Rijomba. She too clearly confesses her love to him that very day:

> I have considered you as my husband right from my childhood ... I would prefer to be your wife than to be an Ane (*Saba Kata Manuh*: 98)

That very moment Darge assures her that he is not going to marry anyone else, if he can not marry her (*Saba Kata Manuh*: 99). Darge's love for Rijomba finds expression at the time of dismembering a deadbody which he initially thinks is that of Rijomba:

> I'll carry your memories with me till I die. In the next life we will take birth together and you will wait for me. (*Saba Kata Manuh*: 139)

After that though he presumes her to be dead, his love for her never decreases. While he meets Noraljom Ane Sangge, her face reminds him of Rijomba. And he can not help trying to meet her again and again. Therefore, though extremely surprised with Ane Sangge's revelation of her real identity as Rijomba, Darge is glad to meet her again:

> I have hoped so...I have hoped that Ane Sangge is my Rijomba ... the same face, the same look, even the same tears, the same way of crying ... O my Rijomba! ... So you are my Rijomba who is alive! (*Saba Kata Manuh*: 160).

The great Rimpoche cries in front of Darge and begs him to sit beside her, "I might be an Ane Rimpoche for others, not for you.... I am Rijomba for you. Don't leave me, please sit beside me" (*Saba Kata Manuh*: 160). Ao Thampa's desire to wipe off her tears and her confession "I prefer to live as Rijomba than as Ane Sangge" (*Saba Kata Manuh*: 165) definitely tell us about the deep love between each other.

Ane Sangge's love for Darge finds its full expression when we are told about the incident of taking off her clothes after her death when we see her wearing the *todung* presented by Ao Thampa inside her dress (the Rimpoche's grey dress). It is significant to note that their love is neither the traditional affair between two young lovers nor does it have a fairy-tale ending.

Another important concern of the novel is the depiction of the traditional Monpa society. We are also told about the food habits of the Monpas, more particularly with reference to the food such as *jhan, churpi* etc. that is offered by his wife to Ao Thampa. We also come to know about the Losar, the Torgya and the other festivals of the Monpas. There are mentions of many beliefs and traditions of the Monpa community. The very custom upon which the story of the novel is based is drawn from the Monpa society and that allows the writer to introduce a host of rites and rituals from the same society. The Monpas slice the dead bodies into a hundred and eight pieces before throwing it into the river. Ao Thampa is involved in this task.

The man who cuts dead bodies earns salvation by cutting the deadbodies into pieces and then distributing it among the fish in the river. Even the soul of the dead showers his blessings upon the man who cremates his body. Slicing one dead body is as good as going on pilgrimage (*Saba Kata Manuh*: 31)

According to the Monpa belief, Ao Thampa's daughter is invalid due to some kind of sin committed during the time of her birth as her father was busy in dismembering the dead bodies of those pilgrims who died in an accident on their way back from Bangajong (*Saba Kata Manuh*: 40).

Rijomba's fear of any dreadful happenings due to the quivering of her right eye is also a reflection of some traditional belief of the Monpas. The author gives the picture of a traditional Monpa society when describing the help offered by others in the community in building the house of Ao Thampa. The Monpas also believe that if a Rimpoche is born in someone's house, some kind of untoward incident takes place in that house (*Saba Kata Manuh*: 175). The writer tells us about the distribution of parental property amongst the wards in the thirty-fifth chapter of the novel when Ao Thampa, Lama Karchung, Yeshe Khando and their sister Yaku inherit the property of their father. Through the description of all these, the writer tries to vividly depict the Monpa society in this novel.

3

New Historicism in Perspective

"Theory" in the context of literature is used in the sense of general principles, together with a set of distinctive terms and categories that is applied to identifying and analyzing works of literature as well as the criteria (the standards or norms) by which these works and their writers are to be evaluated. Right from the time of Aristotle's *Poetics*, there have been countless innovative literary theories which have been revised, countered, and reviewed. In the 20th century, "theory" became an umbrella term for a variety of scholarly approaches to reading texts, most of which were informed by various strands of Western philosophy.

One of the fundamental questions of literary theory is "What is literature?", though many contemporary theorists and literary scholars believe either that the term "literature" is undefinable or that it can refer to a specific use of language. Specific theories are distinguished not only by their methods and philosophy, but even by how they define a "text." It is possible to use the word "text" in a narrow sense to just refer to any of the books belonging to the Western literary canon. But the principles and methods of literary theory have been applied to non-fiction, popular fiction, film, historical documents, law, advertisement etc., and in the related field of cultural studies. In fact, it is possible to treat cultural events like fashion, football riots, etc. as "texts" to be interpreted in the field of cultural studies. Taken broadly, then, literary theory can be thought of as the general theory of interpretation.

Different schools of literary theory have come up with their own answers for the questions "What is literature?" and "What is a text?". For the New Critics literature is a bearer of multiple meanings and they isolate the text as a solid, immutable entity, shielded from any external influences. They believe that "the proper concern of literary criticism is not with the external circumstances or effects of a work, but with a detailed consideration of the work itself" (Abrams, 1993: 246).

Marxist literary criticism seeks to look at literature within its social, economic, and historical context and to understand how the ideas advanced in the work of literature relate to the ideals and values that circulated in the society of the time. It seeks to understand how ideology works in literature to mask social contradictions such as those between economic groups (Ryan, 2002: 53-54). Structuralistic literary criticism views literature as a second-order system that uses the first-order structural system of language as its medium, and is itself to be analyzed primarily on the model of linguistic theory (Abrams, 1993: 280). In post-structuralism, the human search for a truth, a foundation, a 'full presence' is contrasted with the affirmation of 'freeplay' as an ultimate (Blamires, 2000: 363).

Literary theory in general and twentieth century literary theory in particular has very often fluctuated between extremes. For instance, on the one hand we have had the adherents of New Criticism who were emphatic in their advocacy of close reading and attention to texts themselves, and their rejection of criticism based on extra-textual sources and on the other hand we have New Historicism as a mode of literary study whose proponents contest the formalism of New Criticism and the critical beliefs that the latter followed.

In fact, in many ways New Historicism occurred as a response to New Criticism's tendency to treat works of literature in a historical vacuum, as if a poem or novel had no relation to its historical context. So instead of focusing on dealing with a text in isolation from its historical context as New Critics (1940–60) do, New Historicists attend primarily to the historical and cultural conditions of its production and also of its latter critical interpretations and evaluations. There was "a strong 'anti-historical bias' in the New Criticism, mainly because in trying to define the discipline of literary criticism, it was very self-consciously working against what it saw as a dominant historicist approach to literature" (Waugh, 2006: 172). New Critics focused on the text itself ignoring the historical contexts within which literary works are written and read.

New Historicists acknowledge the importance of the literary text, but they also analyze the text with an eye to history. Abandoning certain critical myths such as the aesthetically autonomous work, the formally complete literary icon, the artistic genius at the origin, and the complete or whole reading, Stephen Greenblatt, a leading proponent of New Historicism, suggests instead that criticism concern itself with the collective beliefs, social practices, and cultural discourses that shape a particular work (Ryan, 2002: 129). Stephen Greenblatt's brilliant studies of the Renaissance have established him as the major figure commonly associated with New Historicism. Indeed, his influence meant that New Historicism first gained popularity among Renaissance scholars, many of whom were directly inspired by Greenblatt's ideas and anecdotal approach (Felluga, 2003). This fascination with history and the minute details of culture soon caught on among scholars working in other historical periods, leading to the increasing popularity of culturally and historically oriented studies of literature.

The 1980s witnessed the emergence of the new movement of New Historicism in Anglo-American literary scholarship, as a result of a raising clamour for a return to historical scholarship in the academic study of literature by many scholars in English and American universities - Frederick Crews, George Watson, Ed Hirsch Jr., Frederic Jameson, Terry Eagleton and Frank Lentricchia (Ryan, 2002: 128-29). They believed that the historical nature of a literary work had been neglected over the past half-century of Anglo-American criticism. The time had come to move beyond the narrowly formalistic or text-only approach to literature. A new historical approach was needed and in the course of events, this new movement arose to meet the demand:

> What is called New Historicism is a turn towads sociological, political, and historical criticism, indebted to structuralism and poststructuralism, particularly Foucault, manifested first during the late 1970s in the writings of Greenblatt, Lentricchia, Said and others (Leitch, 1992: 164).

New Historicism developed during the 1980s. As already said, the early 1980s was the prime time for a revolution against texualism after the euphoria following the advent of New Criticism having passed over and it occurred at the University of California at Berkeley. Here, a group of young scholars - Stephen Greenblatt, Catherine Gallagher, D.A. Miller and others like Louis Montrose at U.C. San Diego - launched a "return to history" in literary scholarship (Ryan, 2002: 128-29)

The term 'New Historicism' was coined by the American critic Stephen Greenblatt whose book *Renaissance Self-Fashioning: from More to Shakespeare* (1980) is usually regarded as its beginning. (Barry, 2002: 172).

New Historicist critics are rediscovering the value of history in literary analysis. The historical criticism being practised in the 1980s, however, was not the same as the kind of historical criticism that was practised in the 1930s and 1940s. The most distinctive thing in the new mode of historical study, marking it as poststructural, is mainly the result of concepts and procedures that it has assimilated from various recent theories of literary and other writings. It is influenced by the poststructuralist and reader-response theory of the 1970s, as well as by the thinking of feminist cultural and Marxist critics whose work was also 'new' in the 1980s:

> New Historicism has a portmanteau quality. It brackets together literature, ethnography, anthropology, art, history and other disciplines and sciences, hard and soft (Veeser, 1989: xi).

A number of different approaches to history and culture often get lumped together under the category of "New Historicism". The sheer number of books in historical and cultural studies that have appeared since the early 1990s, and the dominance of the still-larger umbrella term, Cultural Studies, makes the cordoning off of a group of critics as "New Historicists" difficult. The effort to do so is certainly not helped by the fact that some of the most prominent New Historicists, like Stephen Greenblatt and Alan Liu, either reject or critique the very term, "New Historicism" (Felluga, 2003). Nonetheless, this critical school and those scholars commonly associated with the school have been hugely influential on scholarship of the last decade.

The New Historicist critics define history more broadly than their predecessors did. New Historicist critics believe that history itself is not a set of fixed, objective facts but, like the literature with which it interacts, a text which needs to be interpreted. New Historicism acknowledges the fact that all historic views, accounts, and facts they use contain bias. Unlike previous historical criticism, which limited itself to simply demonstrating how a work was reflective of its time, New Historicism evaluates how the work is influenced by the time in which it was produced. It also examines the social sphere in which the author moved, the psychological background of the author, the books and theories that may have influenced the author, and any other factors which influenced the work of art.

New Historicism is a method based on the parallel reading of literary and non-literary texts. A major strength of the New Historicist movement lies in the way it has acknowledged and elucidated the texuality of history and the intertexual relation between canonical literary texts and other kind of cultural data (Wayne, 1991: 801). It simply refuses to 'privilege' the literary texts. It emphasizes and practices a mode of study in which literary and non-literary texts are given equal weight and constantly inform or interrogate each other. This equal importance is best suggested in the definition of New Historicism offered by Louis Montrose who defines it as a reciprocal concern with "the Historicity of Texts and the Texuality of History" (1989: 23). This implies that these critics rethink history, the text, and the interaction between the two. All texts may be examined for their historicity, just as any historical phenomenon can be analysed much as one would a literary text. The New Historicists take this latter precept much further by examining the import of what might be considered rather minor and unimportant daily facts and events, e.g. the cultural and socio-political import of miniature portraiture in Renaissance England.

According to Stephen Greenblatt, New Historicism involves "an intensified willingness to read all of the textual traces of the past with the attention traditionally conferred only on literary texts" (quoted in Barry, 2002:172). Talking about the relationship between literature and culture, Greenblatt says, "... if an exploration of a particular culture will lead to a heightened understanding of a work of literature produced within that culture, so too a careful reading of a work of literature will lead to a heightened understanding of the culture within which it was produced" (1995: 227). History was no longer what it used to be - a background of ideas or a field of empirical facts. New Historicists view history as one that emphasized the role of representation and discourse in social life.

The New Historicists are less fact and event oriented than historical critics used to be, perhaps because they have come to wonder whether the truth about what really happened can ever be purely or objectively known. They are less likely to see history as linear and progressive, as something developing toward the present, and they are also less likely to think of it in terms of specific eras. Hence they are unlikely to suggest that a literary text has a single or easily identifiable historical context.

Though it could be argued that concern with history was already evident in earlier forms of historicism, there is one further major respect in which the latest type of historicism is distinguishable from the old:

> In most of the works identified as New Historicist one senses an absence of a kind of conviction that governed earlier types of historicist criticism (Wayne, 1991: 794).

The earlier types of historicism entailed the conviction that art or literature either reflected or expressed the human subject, however various the nature of the subject might be. According to Frederic Jameson, in the earlier types of historicism, a poetic art was "understood to express, to illustrate or to dramatize a stage in the history of consciousness or of human praxis, whether of an individual or collective kind" (quoted in ibid. 1991). The traditional historicism adhered to such narratives that governed the critic's interpretation of specific narrative or poetic acts. It is the absence of such a conviction that characterizes New Historicism. Though less confident than earlier cultural historians about the precise meaning of 'history', the New Historicists are more self-conscious in acknowledging that the historiographer is never simply describing or retrieving the past but is engaged in constructing a past. The major difference between the 'new' and the 'old' historicism lies in the practices of giving 'equal weight' to literary and non-literary material. (Barry, 2002: 174). Another important difference between old and New Historicism can be seen in the latter's interest in history as represented and recorded in written documents.

New Historicism is not simply "a return to an earlier kind of literary scholarship, for the views and practices of the New Historicists differ markedly from those of former scholars who had used political and intellectual history as a "background" to account for the characteristic subject-matter of literature at a particular time and place" (Abrams, 1993: 248).

The New Historicists combat formalists by dragging cultural/historical considerations to the center stage of literary analysis. In his essay 'Towards a Poetics of Culture' Greenblatt says: "I shall try if not to define the New Historicism, at least to situate it as a practice, a practice rather than a doctrine since as far as I can tell (and I should be the one to know) it's no doctrine at all" (1989: 1). It is important that the term 'New Historicism' does not pertain to a school or to a coherent theory of cultural history. Rather it designates a fairly diverse body of scholarship with some common attributes. This is a type of criticism that deals principally with the importance of local, political and social contexts for the understanding of literary texts:

> Greenblatt's own *Renaissance Self-Fashioning* (1980) and the journal *Representation* founded by Gallagher, Walter Benn Michaels, Greenblatt and others consolidated the New Historicism not as a doctrine but as a set of themes, preoccupations and attitudes (Veeser, 1989: xiii).

New Historicism has borrowed these themes, preoccupations and attitudes from various recent theories and other writings. In the introduction to *Renaissance Self-fashioning* (1980) Stephen Greenblatt called his project a 'cultural poetics' (1988). He has returned to this term in the introductory chapter of his *Shakespearean Negotiations: The Circulation of Social Energy in Renaissance England* (ibid. 1989: 32). He lays out some of the theoretical bases of the project of cultural poetics in these books. Drawing on Michel Foucault, the *Annales* School of social history, the cultural anthropology of Clifford Geertz and deconstruction, Greenblatt argues for a "cultural poetics" that would study the social and cultural negotiations, transactions and exchanges that go into the making of a literary work (Ryan, 2002: 129). In some ways, Greenblatt seems to be echoing the words of the anthropologist Edward B. Tylor (1871):

> Culture or civilization taken in its wide ethnographic sense, is that complex whole which includes knowledge, belief, art, morals, law, custom, and other capabilities and habits acquired by man as a member of society (quoted in Greenblatt, 1995: 225).

Stephen Greenblatt begins his most theoretical statement about New Historicism, 'Towards a Poetics of Culture', by stating that his methodology is, at best a "practice" rather than a "doctrine":

> One of the peculiar characteristics of the 'New Historicism' in literary studies is precisely how unresolved and in some ways disingenuous it has been—I have been—about the relation to literary theory (1989: 1).

New Historicists remind us that it is treacherous to reconstruct the past as it really was—rather than as we have been conditioned by our own place and time to believe that it was. And they know that the job is impossible for those who are unaware of that difficulty, insensitive to the bent or bias of their own historical vantage point. Thus, when New Historicist critics describe a historical change, they are highly conscious of (and even likely to discuss) the theory of historical change that informs their account (Murfin and Ray, 1998: 240).

According to the New Historicists, a text, whether literary or historical, is a discourse which, although it may seem to present, or reflect, an external reality, in fact consists of representations, that is the verbal formations which are the 'ideological products' or 'cultural constructs' of a particular era. Far from being divorced from the world, literary production is itself a form of social practice: texts do not merely reflect social reality but create it:

> Under the influence of this essentially anthropological perspective, recent literary historians have sought to replace literary texts within the larger cultural formations of which they were originally a part (Patterson, 1995: 260).

These cultural and ideological representations in texts serve mainly to reproduce, confirm and propagate the power-structures of domination and subordination which characterize a given society. Therefore, representations in literary texts do not reflect reality but are "concretized" forms of ideology. Moreover, a literary work is less the product of its author's imagination than the social circumstances of its creation. The New Historicists emphasize the importance of ideology, the political disposition, unknown to an author himself that governs his work. New Historicists like Stephen Greenblatt "have sought to rescue biographical criticism by renewed examination of the complex sociohistorical forces that constitute subjectivity" (Leitch, 1992: 26).

Ideology is not a set of doctrines; it signifies the way men live out their roles in class-society, the values, ideas and images which tie them to their social functions and so prevent them from a true knowledge of society as a whole. The concept of ideology developed in Marxism because Marxism always sought to be not just a narrow political theory but a more comprehensive theory that could explain the complex and important relations among the political, economic and cultural elements in specific societies. Ideology became the term through which Marxists "tried to articulate, in various ways, the relation between the realm of culture (including, but not limited to "ideas") and the realm of political economy (including "production")" (Kavanagh, 1995: 307).

Marxist thinker Louis Althusser's concept of ideology (in 'Ideology and the State', 1992) had a powerful impact on the beliefs and concepts that framed the New Historicist approach. Louis Althusser has reworked the concept of ideology in the light of the more complex notion of subject-formation given by psychoanalysis. In this framework, 'ideology' designates a rich "system of representations" worked up in specific material practices, which helps form individuals into social subjects who "freely" internalize an appropriate picture of their social world and their place in it (Kavanagh, 1995: 310). 'Ideology' manifests itself in different ways in the discourse of each of the semi-autonomous institutions of an era, and 'ideology' operates to position its readers, 'subjects' them - that is, subordinates them to the interest of the ruling classes. 'Ideology' signifies 'the imaginary ways in which men experience the real world, which is of course the kind of experience literature gives us too - what it feels like to live in particular conditions, rather than a conceptual analysis of those conditions' (Eagleton, 2002: 16). The New Historicists are concerned with the recovery of the 'ideology' which 'gave birth to the text', and which the text in turn helped to disseminate throughout a culture:

> The contribution of the New Historicism has been to identify a third alternative in which the very antagonism between literature and ideology becomes, in specific historical environments, a powerful and socially functional mode of constructing subjectivity (Gallagher, 1989: 44).

This dimension of critical interpretation has been neglected by traditional historicists, not because the concept of 'enabling presumption' of ideology was unavailable to them until recently, but it had never been widely attempted because literary texts themselves suppress the means by which they construct ideology. Only a New Historicist approach, treating text as one element in the ideology of an age, can hope to lay them bare.

In so far as New Historicism attempts to understand the relationship between a text and the political, social and economic circumstances in which it originated, it is quite close to Marxism. But whereas Marxism tends to see literature as part of a 'superstructure' in which economic 'base' manifests itself, New Historicists tend to take a more nuanced view of power, seeing it not exclusively as class related but extending throughout society. This view derives primarily from Michel Foucault (Abrams, 1993: 249).

A number of New Historicist critics owe a great debt to Foucault, a French philosophical historian who tried to redefine the boundaries of historical inquiry bringing together incidents and phenomena from areas normally seen as unconnected. The New Historicists make use of aspects of Foucault's idea of 'social structures' as determined by 'dominant discursive practices'. They focus attention on issues of state power and how it is maintained, on patriarchal structures and their perpetuation, and on the process of colonization with its accompanying mindset:

> In a now-famous play on words, we might say that "power" through its discursive and institutional relays "subjects" us: that is, it makes us into "subjects", and it "subjects" us to the rule of the dominant disciplines which are empowered in our society and which regulate its possibilities for human freedom - that is, it "subjugates" us (Bove, 1995: 58).

Foucault in his book *Nietzsche, Genealogy, History* observed that while the "traditional historians" emphasize the difference between the present and the valorized past, the "effective historian" collapses the distance that separated the past from the present (quoted in Panja, 2002: 192). He observed that history is not a battle or a treaty but a "reversal of the relationship of forces"(ibid. 2002). He refused to see history as an evolutionary process, a continuous development towards the present, or an abstraction or an idea:

> No historical event, according to Foucault has a single cause rather each event is tied into a vast web of economic, social and political factors. Foucault urged historians to be aware that they are themselves historically "situated" making it difficult to see present cultural practices critically and extremely difficult to enter by gone ages (Murfin and Ray, 1998: 243).

Foucault saw history in terms of power and he viewed power not simply as a repressive force or a tool of conspiracy but rather as a complex of forces that produces what happens. Stephen Greenblatt has written about the influence of Foucault and his thoughts on him in his essay, 'Towards a Poetics of Culture':

> Certainly the presence of Michel Foucault on the Berkeley campus for extended visits during the last five or six years of his life, and more generally the influence in American and European (and especially French) anthropological and social theorists, has helped to shape my own literary critical practice (1989: 1).

Foucault's concept of power and history is best reflected in Judith Lowder Newton's voice in his essay 'History as Usual? Feminism and the New Historicism':

> There is the notion that 'history' is best told as a story of power relations and struggle, a story that is contradictory, heterogeneous, fragmented, the history is a tale of many voices and forms of power exercised by the weak and the marginal as well as the dominant and strong (Newton, 1989: 152).

Incorporated into the New Historicists' preference for the textual record of the past is the influence of deconstruction. New Historicism accepts Derrida's view that there is nothing outside the text, in the special sense that everything about the past is only available to us in a texualised form: simply every fact of reality is texualised

(Barry, 2002: 175). The concept of deconstructive criticism that all texts involve modes of signification that was against each other, merged with Mikhail Bakhtin's (ibid. 2002) concept of the dialogic nature of many literary texts, in the sense that they present a number of independent and often conflicting voices. According to the New Historicists the relationship between literature and history is one of circulation, exchange and negotiation rather than of reference or reflection. There is nc single historical discourse of a period; instead the critic must trace out the multiple and complex interconnections of histories that make up an age. The use of terms such as 'circulation', 'negotiation', 'exchange',' transaction' etc are also the result of the influence of deconstruction and Jacques Derrida.

Stephen Greenblatt has said:

> My use of the term 'circulation' here is influenced by the work of Jacques Derrida (1989: 8).

Recent developments in cultural anthropology, especially Clifford Geertz views that a culture is constituted by distinctive sets of signifying systems and his use of 'thick description' has a powerful influence on New Historicism. "Thick description" is the close analysis or reading of a particular social production so as to recover the meaning it has for the people involved in it, as well as to discover the patterns of conventions, codes and modes of thinking that invest the cultural item with those meanings (Abrams, 1993: 249). Clifford Geertz's friend Lawrence Stone, the historian, describes "thick description" as: "A close and well-informed look at seemingly trivial acts, events, symbols, gestures, patterns of speech or behaviour which then can be made to reveal whole systems of thought...and...problems of kinship, lineage or community structures" (quoted in Panja, 2002: 191). It is a term used to describe the seemingly insignificant details present in any cultural practice. By focusing on these details, one can then reveal the inherent contradictory forces at work within culture.

Anthropology, especially anthropology as practised by Victor Turner and Clifford Geertz is found particularly useful in the New Historicist analysis of literature. Geertz has influenced many New Historicist critics (Murfin and Ray, 1998: 240). It is largely due to Geertz's influence that New Historicists have asserted that literature is not distinct from the history that is relevant to it. New Historicists reject the separation of artistic works from their creators and audiences. They have used what Geertz calls 'thick description'. They have erased the line dividing historical and literary materials. Hence, when New Historicist critics describe a historical change, they are highly conscious of the theory of historical change that informs their account. They know that the changes they notice are ones that their (historically-determined) theory allows or helps them to see and describe.

It has been observed that New Historicism does not pertain to a school or to a coherent theory of cultural history. Instead it designates a fairly diverse body of scholarship with some common attributes. Varied influences of different mode of criticism and critics and of cultural critique, of anthropology and Marxism, of Feminism and Post structuralism have definitely framed New Historicism as a mode of study with considerable diversity and disagreement among individual exponents of it. However, a few salient features can be summed up within these overall perspectives which are found to occur frequently in their writings, sometimes in an extreme and sometimes in a qualified form. These are some methodological features which can be found to characterize all these approaches which have been assimilated in New Historicist approach. For one, New Historicists believe that literature is not independent of economic, social and political conditions and therefore, not subject to timeless criteria of artistic value. In a given historical moment, different modes of discourse such as law, theology, moral philosophy, literature, art, architecture etc are not autonomous, but the constitution of a given cultural field and of some ideological.codes that order all discourse in that particular culture. Therefore, a literary text which is simply one among many kinds of texts - religious, philosophical, legal or other - is always subject to the particular conditions of a time and place and it has neither unique status nor special privilege.

New Historicists tend to refuse hierarchies and dichotomies among texts of different kinds (canonical/ non-canonical; high culture/mass culture; documents/ fictions). One prominent tendency is to subvert the hierarchical distinctions between 'high art' and 'high literature' and the traditionally 'lower' forms that appeal to a much larger body of consumers (Abrams, 1993: 254). They believe that the boundary between literary and non literary texts is a false one.

History for the New Historicists is not a homogenous and stable pattern of facts and events which can be used as the "background" to explain the literature or which can be said simply to reflect. According to the New Historicist, a literary text is an interactive component within the network of institutions, beliefs and cultural power-relations, practices and products which constitute what we call history.

Revival of interest in the history of rhetoric, though from a critical (rather than a merely descriptive) perspective is an important feature of New Historicism. New Historicists always talk about the systematic reading of the wider cultural field by means of an attention to rhetorical devices and strategies.

New Historicists have shifted interest from ideas to power relationships as the fundamental units for analysis and interpretation in cultural history. Such a shift, on the other hand leads to focus on issues and patronage, patriarchal authority and its legitimation, the role of culture in the formation of the modern nation state. The governing assumption is that culture is an active force in history as discourse and representation rather than merely reflecting or expressing it. The concepts, themes and procedures of

New Historicist criticism took shape in the late 1970s and early 1980s and in the course of 1980s, the view points and practices of New Historicism spread rapidly to all periods of literary study.

New Historicism as described already is mainly the result of concepts and procedures that it has assimilated from various recent theories of literary and other writings. At the same time, many other literary and critical theories were also influenced by this new mode of criticism. Some feminist critics adapted this new historical mode to their concerns by stressing the role of masculine power-structures in forming ideological and cultural constructs (Abrams, 1993: 252). Some others interested in black and other ethnic literature stressed the role of culture-formations dominated by white Europeans in suppressing, marginalizing, or distorting the achievements of non-white and non-European peoples.

Stephen Greenblatt, for instance, reads a selection from Thomas Harriott's 'A Brief and True Report of the New Found land of Virginia' written in 1588, as a representative discourse of the English colonizers of America (Ryan, 2002: 129). He asserts that Harriott (1588) tests the English power structure by recording the counter-voices of the American Indians who are being appropriated and oppressed by that power. Greenblatt then identifies similar modes of power discourse and counter-discourse in the dialogues in Shakespeare's *The Tempest* between Prospero the appropriator and Caliban the expropriated native of his island. He reads the Henry plays as revelation of the degree to which princely power is based on predation, calculation, deceit and hypocrisy, at the same time, finds that these plays do not scruple to record the dissonant and subversive voices of Falstaff and other representatives of Elizabethan sub-cultures (Abrams, 1993: 252-253).

Marxist critic, Raymond Williams is a powerful influence on the New Historicist beliefs and concepts (ibid. 1993: 253). The Marxist orientation of New Historicism owes largely to Raymond Williams and his term "Cultural

Materialism". A number of British scholars have adopted this term to indicate the Marxist orientation of their mode of New Historicism. They insist that the "texuality" of history, a culture and its literary products are always to an important degree, conditioned by the real material forces and relations of production in that historical era. Both New Historicists and Cultural Materialists are interested in recovering lost histories and in exploring mechanisms of repression and subjugation. Though each of the schools practises different kinds of history, New Historicists tend to draw on the disciplines of political science and anthropology given their interest in governments, institutions, and culture, while Cultural Materialists tend to rely on economics and sociology given their interest in class, economics, and commodification (Felluga, 2003). New Historicists are, like the Cultural Materialists, interested in questions of circulation, negotiation, profit and exchange, i.e. how activities that purport to be above the market (including literature) are in fact informed by the values of that market. However, New Historicists take this position further by then claiming that all cultural activities may be considered as equally important texts for historical analysis: contemporary trials of hermaphrodites or the intricacies of map-making may inform a Shakespeare play as much as, say, Shakespeare's literary precursors (ibid. 2003). New Historicism is also more specifically concerned with questions of power and culture (especially the messy commingling of the social and the cultural or of the supposedly autonomous self and the cultural/ political institutions that in fact produce that self).

Cultural Studies is another recent but very rapidly growing cross-disciplinary enterprise related to New Historicism. The name designates the critical analysis of both the production and reception of all forms of cultural institutions, processes and products; among these, literature counts as merely one of many types of "symbolic constructions" (Abrams, 1993: 253).

Within the ranks of the New Historicism, literature is considered to be one of the social forces that contribute to the making of individuals; it acts as a form of social control. The central task of the New Historicism is firstly to call into question the traditional view of literature as an autonomous realm of discourse with its own problems, forms, principles, activities, etc.; and secondly to dissolve the literary text into the social and political context from which it issued.

New Historicists try to relocate the literary text among the other, traditionally nonliterary cultural practices of an age. The representation of character in the nineteenth-century novel, for instance, is said to be bound up with contemporary debates over parliamentary representation; or, Iago's plot against Othello is described as typical of Elizabethan attempts to deny the otherness of subject peoples (Myers, 1988: 30). But the larger purpose of New Historicist inquiry is the reconstruction of the actual (as opposed to the "represented") relations in which people lived during a particular time. For example, in one of the most widely read essays by a New Historicist, Louis Adrian Montrose interprets *A Midsummer Night's Dream* as an ideological attempt to comprehend the power of Queen Elizabeth - to make sense of it and place it safely within bounds—while simultaneously upholding the authority of males within Elizabethan culture (quoted in ibid. 1988). By citing a variety of contemporary writing (in order to reinstate the "discursive practices" of the age), Montrose demonstrates the Elizabethans' ambivalence toward their queen: abiding respect mixed with a dark desire to master her sexually. In this context, *A Midsummer Night's Dream* is reread as a fable of the restoration of male governance. Mothers are significantly excluded from the *dramatis personae* of the play, just as the danger of matriarchy (with which the Elizabethans flirted in their fascination with the myth of the Amazons) was quietly suppressed by the celebration of Elizabeth's virginity. The very real possibility that power might actually be passed from mother to daughter was concealed from women of the age by such cultural productions as Shakespeare's play, in which Elizabeth was a willing collaborator as much by her decision to remain unwed and "barren" as by her "cultural presence" within the play.

It is in this sense that works of literature such as *A Midsummer Night's Dream* are "representations" of the culture from which they emerge. They are the emanations, the active agents, of the culture's ideology. Literary works are both what a culture produces as well as what reproduces the ideology. When a New Historicist looks at a novel by Jane Austen, he/she would observe that it is confined to a very limited sphere of society, namely the landed gentry. While the New Historicist may praise the novel, he or she will also duly note that the servant class is completely marginalized in Austen's work. Austen asserts the pre-eminence of the landed gentry above any other class of society, and is quite critical of those who marry "beneath" their social status. The critic might then evaluate why Austen would display this prejudice, giving information about her background, the books she had read, events in her life that may have influenced her, and her own choices in regards to marriage. Austen is, in a way, against her own work, which suggests power may be purchased through good marriages, since Austen never married. In fact, Austen's life stands outside her own espoused theories in literature, because she was a female novelist, gaining power through her work rather than through marriage. One practicing New Historicism would note such a dichotomy. The fact that Austen wrote about a society to which she only marginally belonged can enhance the reading of her work. It might also help make sense of her broad critiques of "civilized" society.

In view of this, it becomes impetrative to situate Lummer Dai and Yeshe Dorjee Thongchi's narratives in the time/age to which it belongs- the first part of the 20th century. In the next chapter I will look into the collective beliefs, social practices, and cultural discourses that must have shaped Dai and Thongchi's views as expressed in their novels and would, more importantly help to explain the writers' concern in depicting the life and society of the characters of that particular time and age.

4

Ethnic Worlds in Lummer Dai

In the previous chapter I discussed how the New Historicists try to establish the inter-relation between literature and culture of a period. M.N. Srinivas says, "Fiction [...] can be a valuable source of information and insight into culture" (1998: 2528). New Historicism attends primarily to the historical and cultural conditions of the production of a text (in place of dealing with a text in isolation). Lummer Dai's novels are not merely a reflection of some reality but 'representations' (the key word for the New Historicists) - the verbal formulations that are the products of his society, of the Arunachalee tribal society of the first half of the twentieth society, particularly the Adi society. As Michael Ryan writes,

> According to [N]ew [H]istoricists, the relationship between literature and history is one of circulation, exchange and negotiation rather than of reference or reflection (2002:130).

Stephen Greenblatt, a prominent New Historicist, argues for a 'cultural poetics' (Greenblatt,1990:3) that would study the social and cultural dialogue, transactions and exchanges that go into the making of a literary work. He suggests that New Historicism concerns itself with the collective beliefs, social practices and cultural discourses that shape a particular work.

In other words, in the aftermath of New Historicism it would be imprudent to assume that the literary text exists in vacuum and is created out of nothingness. Every page, every sentence of a literary text has its origin in the cultural discourses and social practices and ideologies of a society. In view of this, it is vitally important to situate Lummer Dai's novels in the period, culture and society to which they belong - that is the tribal society of Aruanchal Pradesh in general and the Adi Society in particular. Terry Eagleton says:

> The author is essentially a producer who works up certain given materials into a new product. The author does not make the materials with which he works: forms, values, myths, symbols, ideologies come to him already worked upon as the worker in a car-assembly plant fashions his product from already processed materials (Eagleton, 2002: 64).

Lummer Dai wrote with the avowed intention of making the customs and rituals of his society known to the outside world:

> There is hardly anyone to try and portray the ways of life, social customs and rituals of the society of Arunachal in work of literature. Hence, I try my best to write about our society (Ajanta Dutta: *Aaji*, 3rd Feb, 2002).

However, it would be for us to analyze if Dai is succesful in retaining his autonomy as an author, who as much as anyone else, is a "subject" produced and positioned by the play of power and ideology within the society from which he comes (Abrams, 1998: 251).

In Lummer Dai's *Paharar Xile Xile* (1961) the readers are introduced to the Adi society of Arunachal Pradesh (Biren Bhattacharya, Introduction to *Paharar Xile Xile*). The novelist says in the preface to the novel that the story is based on the love affair between Bati and Jalam, representatives of two separate classes of the society - Bati from the lower class or the class of slave and Jalam, the male protagonist from the upper class. In this context, Jitanjali Borpujari observes: "Bati's tears are the tears of many slaves' daughters who had to live under restriction though freed from slavery" (Borpujari, 2000:112).

The writer makes an interesting comment in the introduction to the novel that allows us to read the novel from a New Historicist approach:

> The character of Bati is a completely imaginative one; such a character is not a rare one in the traditional society - a woman of the remote hill deprived of the light of civilization. On the other hand, the system of slavery in the Adi society though completely done away with now-a-days, one born in a slave's womb is still considered as belonging to the lowest class of the society (*Paharar Xile Xile*: Preface).

Markist thinker Louis Althusser's concept of ideology can be used to analyse Bati's hesitation in accepting Jalam's proposal for marriage. Even if she loves Jalam she is not convinced or happy with his proposal: "I know you love me, but if you marry me, it would hurt your parents, because I am the daughter of a slave. You may love me but don't talk of marrying me" (*Paharar Xile Xile:* 6).

It is the ideology that Bati has indoctrinated from her childhood that comes in the way of her happiness, i.e., to accept Jalam's proposal of marriage, Ideology, in fact, "interpellates the individual to take up a pre-established 'subject - position'" (Abrams, 1993: 244). For Terry Eagleton, ideology is not just "a set of doctrines; it signifies the way men live out their roles in class-society, the values, ideas, and images which tie them to their social functions and so prevent them from a true knowledge of society as a whole" (Eagleton, 2002: 15). Ideology interpellates Bati to take up the pre-established subject position of a lower class woman, a slave, which does not allow her to go beyond certain "limits" that society has imposed upon her. This ideology makes her believe -

> What for would I marry; if your parents refuse to eat anything I serve? The duty of a daughter-in-law is to look after her in-laws, to serve her husband. If that cannot happen, there is no sense in being married (*Paharar Xile Xile:* 6).

The sensitive issue of the practice of slavery is an important theme of the novel *Paharar Xile Xile.* According to novelist, there is still a section of 'lower class' people living as *bandi* (slave) in the Adi society:

> Just till a few days ago, there was the system of slavery in the Adi society. The slaves were bought. Even now a section of lower class people are considered as slave, though they are given liberty today (*Paharar Xile Xile:* 15).

Similarly, Jalam's father Bijong gets angry when he hears about his son's relationship with Bati and exclaims in anger: "I do not want a slave's daughter as my daughter-in-law. I do not want even a drop of slave's blood in my family" (*Paharar Xile Xile:* 16). Dai, being an insider of the Adi society, must have had a lot of experiences relating to slavery. The conflict within the characters in the novel is due to the strong faith in the system of slavery.

Tamo Mibang, an Adi intellectual, says about slavery: "Slavery existed in the Adi society, but practically there was no casteism or racialism though social grades existed which was a common phenomenon in human society" (2000: 423). Tai Nyori also accepts the prevalence of slavery in the Adi society of the past but says that it was maintained for "economic and social necessities" (1993: 211). Lummer Dai takes up slavery as one of the central issues of the novel. He tries to expose this social system of slavery for what it is. It was not as easy as it sounds for 'slavery' is a very 'sensitive' issue for intellectuals of

Arunachal, as I was informed in a personal interaction with Tamo Mibang that at present the educated Adi people even hesitate to utter the word 'slave'. Obviously, it required the comprehensive soul of Lummer Dai to talk about this issue frankly. In this we find Lummer Dai being able to retain his autonomy as an author.

Dream plays a very important role in the tribal mind and this is reflected in Dai's novel *Paharar Xile Xile*. Dai explains the belief associated with dream (in the footnote):

> The Adis believe that if someone tries to cross a river or climb a hill in a dream, it causes danger in real life. To overcome the danger, he/she has to cross the river or to climb the hill successfully (*Paharar Xile Xile:* 30)

In the novel, the male protagonist Jalam sees dreams when he is sleeping. Before he goes to sleep he is thinking about Bati and the difficulties to be faced by him in making her his wife. In the dream he is ready to fight with the strong current of a river and he gets down there, but he cannot swim and is swept away by the river (*Paharar Xile Xile:* 29). Frank Lentricchia says in *Foucault's Legacy: A New Historicism:*

> For a New Historicist, literature is no cool reflection on a "back ground" of stable and unified historical fact. It is at once part of the "fact" itself and what gives shape to what we know as the fact (1989: 234).

New Historicists focus attention (within both text and co-text) on issues of patriarchal structures and their perpetuation. Peter Barry (1995) has mentioned Louis Montrose's essay "*A Midsummer Night's Dream* and the Shaping of Fantasies of Elizabethan Culture: Gender, Power, Form" (appeared originally in the American journal *Representations*) to show how patriarchy is maintained even in the society of Elizabeth in spite of the presence of a woman at the pinnacle of power by constantly insisting on Elizabeth's difference from other women. Thus "Elizabeth's rule was not intended to undermine the male hegemony of her culture. Indeed, the emphasis upon her difference from other women may have helped to reinforce it" (Barry, 1995:181).

In *Paharar Xile Xile,* this note of patriarchy is revealed in different circumstances. The women characters, bold and active in other ways, too are not free from the patriarchal male dominated society. Jadi, who is brought to Jalam to stay as his wife expresses her situation in frustration:

> If I do have a husband, why don't I see him, I do not want to be Jalam's wife for name's sake. Let me go and stay with my parents, you can take the bride-price from my second husband (*Paharar Xile Xile*: 35).

The inhuman torture of Bati, particularly when her husband drags her out of the house "like a dead dog" (*Paharar Xile Xile:* 77) when she is ill is a reflection of the kind of male chauvinism prevalent in this society.

Stephen Greenblatt has rightly observed:

> Girls were also the objects of a sustained cultural scrutiny that focussed on the critical passage from the authority of the father or guardian to the authority of the husband. This transition was of the highest structural significance, entailing complex transactions of love, power and material substance, all of which, we may note are simultaneously at issue when Lear demands of his youngest daughter a declaration she is unwilling or unable to give (Greenblatt, 1990: 84).

The character of Bati in *Paharar Xile Xile* is quite similar to that of Cordelia, the youngest daughter of Lear, particularly when she rejects her parents' proposal for going back home:

> I shall not go I had not come here to go back to your hourse. My husband's house is mine. If I had to die, I shall die here. Though you gave me birth, but my husband is more close to me than you (*Paharar Xile Xile:* 78).

It is quite obvious that Bati, true to the tradition of a patriarchal society, has by now transferred all sense of loyalty from her father to her husband. She in fact, asserts her closeness to her husband with greater conviction than Cordelia considering the treatment meted out to her by her husband.

Lummer Dai's second novel *Prithivir Hanhi* (1963) offers us a vivid picture of the Adi Society and Adi social life. According to Borpujari, Dai has shown "the complete history of the Adi society in this novel" (2000: 112). Satyendra Nath Sharma has also said that the novelist is successful in depicting a realistic picture of the tribal life of Arunachal in this novel (Sharma, 1976).

New Historicism presumes that artistic fiction does not imitate human action, it mediates it. That is, fiction is defined as the lens through which a certain portrait of the human experience is brought into focus. And as mediation rather than as imitation of social practices, it can be said to shape rather than to reflect a group of people living in a particular age and their understanding of human experience and potentiality.

The description of the Adi village and the *mosup* in minute details in *Prithivir Hanhi* definitely make us believe that only one who is born and brought up in a society like this would be able to picture it so accurately and realistically. The description of the *mosup* (*Prithivir Hanhi*: 96) and the description of on Adi house (*Prithivir Hanhi*: 117) takes the readers to a typical Adi village, which the novelist must have seen and lived in. We can quote Murfin and Ray in this regard to relate to the New Historicist understanding of this novel:

> New Historicist critics assume that works of literature both influence and are influenced by historical reality and they share a belief that literature both refers and is referred to by things outside itself (Murfin and Ray, 1998: 240).

Lummer Dai talks about the kind of things that are prized possessions for Adis - mithuns, dishes and bowls. He also tells us about their ways of life including descriptions of the preparation and serving of local liquor (*apong*) in festivals to welcome guest. All these are reflections of the actual Adi society from which he has taken "the materials to work with" (Eagleton, 2002: 64).

In *Prithivir Hanhi,* Lummer Dai has portrayed the breakdown of the simple way of the people of these hilly regions of Arunachal with the spread of "mainstream" Indian civilization. The social transformation taking place in the Adi society is an important issue that the novelist has tried to trace. Jagdish Lal Dawar says in this regard:

> The most important issue which agitated the Adi intellectuals was the fear of disintegration of their traditional culture and thereby loss of their identity as a result of the forces of modernisation (2000: 417).

It is perhaps this sense of fear which, as Dawar points out, made Lummer Dai end the novel without providing any solution to certain questions he had raised.

(a) In what direction is the Adi society moving?

(b) What would be the nature of the Adi's cultural identity faced with the intrusion of a new value system and destruction of the old way of life? (ibid. 2000)

Lummer Dai, an Adi intellectual is also perhaps disturbed by such issues due to which he could present the conflict between nature and civilization in *Prithivir Hanhi*. In fact, the conflict between nature and civilization shapes his writing. So, he portrays Gachi, the innocent and honest tribal girl and Lib, the straight forward, innocent, naïve tribal boy as representing nature and

Kardug, the cunning, speculative self-seeking, deceptive person as close to the modern commercial society and as representing civilization.

According to J.L. Dawar (2000) the traditional Adi society has been ungoing a change due to various developmental activities which threatens the traditional way of life. It has led to a crisis of identity in the society, especially among the "intellectuals" - a word he uses to denote "writers, public man, journalists and thinkers" (Dawar, 2000: 409). Lummer Dai fits into this concept of intellectual as a writer, a public relation officer, a journalist (former editor of *Echo of Arunachal*) and obviously, as a thinker. As an intellectual, he was obviously disturbed by the clash of tradition and forces of modernisation. Being always ready to welcome civilization, urbanisation and modernity, he was aware of the debilitating effects of these. It is this awareness that makes him depict the characters of Gachi and Lib - ascribing all the values associated with a simple tribal 'subject' - innocent and naive, honest and straightforward as well as the character of Kardug - a speculative, cunning deceptive person "attributed with all negative features of a commercial society, in other words, of civilization" (ibid. 2000: 417). Kardug takes Lib to Dibrugarh to sell *jim* cloth and not only steals his money but also deserts him. Through this the writer has tried to show the picture of loss of faith and the breakdown of the simple tribal life with commercial expansion. The transformation of the Adi society in reality is clearly captured in Lummer Dai's novel through the incident of the betrayal of a close friend for some material benefit. Such transition in the society and social norms is shown in another situation by Lummer Dai when he makes Liyi take a very strong stand in the *kebang*. We hear a voice of protest, a voice of a mentally strong girl who dares to speak against the age-old pinnacle of power - the *kebang* when she says:

> Is this what you mean by *kebang*? Is *kebang* an institution that makes one shed tears, that snatches

> the food of the poor, that which adds to the suffering and torture of the exploited ...? Is this an institution where opportunists are always the beneficiaries? (*Prithivir Hanhi:* 297).

Sachin Roy says that in real social life of the Adis, no woman is formally allowed to "join a *kebang*" (quoted in Mibang, 2000: 424). Liyi not only makes her presence felt in the *kebang* but also protests against its decision. This is something quite unusual and it tells us about the transformation taking place in the society. This is again an instance of the manner in which Lummer Dai was able to transcend the ideology of his society sufficiently to represent the "objective" reality of his time.

The tribal societies believe in spirits and supernatural entities and the tribal people worship different deities. The Adi society is no exception in this regard. *Prithivir Hanhi* is full of such superstitions and beliefs in supernatural beings. Being born in such a society, the novelist cannot set himself free himself from such beliefs. It is quite possible that he must have dwelt in confusion regarding accepting or rejecting the traditional beliefs of his society. According to Greenblatt, Shakespeare was in doubt whether to follow the catholic rituals or the protestant parctice of abolition of purgatory at the death of his son Hamnet just before his writing of the play *Hamlet* (Gewertz, 2004). The possible dilemma for Shakespeare was whether to risk punishment by praying for the dead son or obey the law.

Here, the dilemma for Lummer Dai is whether to risk a life advocating to an age -old custom or to speak against it for the peace and progress of the society. Dai talks about the belief of associating ill luck with seeing snakes on the way (*Prithivir Hanhi*: 19), the spirit which aggravates wounds if someone calls it small (*Prithivir Hanhi*: 16), the *Bandevata* who showered his blessings on Lib enabling him to kill a deer (*Prithivir Hanhi*: 91), offering rice to satisfy the household deity (*Prithivir Hanhi*: 128). At the same time, Dai depicts the character of Kardug as a cunning, deceptive man, who thinks only about his selfish motive to make money and cheat Lib, his friend.

As an intellectual of the Adi society, Lummer Dai perhaps wanted to welcome the progress of the Adi society in the name of urbanisation and civilization and at the same time, he wanted to preserve the cultural heritage, tradition and local beliefs and rituals. As such, traditional beliefs and rituals of the Adi society do find a place in Dai's novel. But Dai's attitude toward these beliefs is non-committal and it is not difficult to see why.

Motifs help to further traditional beliefs regarding people of a particular society. The motif of the tribal as simple, naive, uncivilised, friendly, helpful is seen in characters like Lib: the reader sees his wish to search for a beggar and give him money, his first experience with an electric light, his first stay in a hotel and his reaction to it. In contrast, another motif that finds an expression in the novel is that of the native as extremely cruel and unkind. This motif finds expression through the character of the fickle-minded Kardug.

It is the ideology indoctrinated from childhood in his society that allows the writer to portray the character of Kardug as a male chauvinist in a patriarchal society who treats woman as a mere commodity:

> I'll get cash 200 rupees or a mithun from your new husband as a compensation just sitting at home. How many people would be lucky enough to acquire such a lot of property just like that without any hard labour?" (*Prithivir Hanhi*: 114).

Woman is seen as a commodity that can be possessed or can be thrown away and can even be exchanged with money or other material things. Having indoctrinated ideas of male supremacy, Kardug even has the courage to physically abuse Liyi in front of all the members of the *kebang*. The *kebang* that enjoys supreme power in the Adi society maintains silence in the face of such abuse of woman. Although Dai does present Liyi as a girl who dares to speak against the crime of a powerful male, he cannot go to the extent of depicting her ultimate victory. The bold and truthful Liyi becomes a victim of male chauvinism and dies in the *kebang*. Of course, this can also be seen as Dai's own loss of faith in the *kebang* as the ultimate source of justice. Dai's doubts regarding the *kebang* as the supreme souce of justice is reflected in Liyi's voice when she is talking to Bankong:

> "Whom will [the *kebang*] believe? Either me or Kardug. It will believe me because I am innocent and the *kebang* is the provider of justice. It might also believe Kardug because he has good links with the socially powerful people of the society. Ultimately the *kebang* will blame me for everything, as it is an institution of injustice and sin. It is a personal institution of the opportunists (*Prithivir Hanhi*: 215).

We can hear the author's voice here. He is perhaps dwelling between the faith and respect for the traditional platform for true judgement on the one hand, and the loss of faith of the new generation in it.

Like his other novels, *Mon aru Mon* (1968) is a novel that includes a variety of traditional beliefs, social customs and rituals. These beliefs are either described by the narrator or observed by any character(s) of the novel. However, it is important to note that the writer introduces the readers to the clash of different characters regarding observance of such beliefs. In fact, the novel expresses a kind of conflict, not only in the minds of the characters' but also in the mind of the writer. The narrator's direct comment, comments through the voice of the characters of the novel, mode of describing the folk beliefs and even the footnotes definitely reveal many things not only about the societies but of the writer and his attachment to the society he is writing about. Dai's sense of affiliation to his roots allows him to include a variety of Adi rituals, beliefs and tradition in his fictional world. Kishore Bhattacharjee says:

> In the north east, where we have very little written records about most of the societies, the texts have greater value (2004: 26).

Dai always keenly observes the response of his characters to the traditional beliefs and customs. It is not that all his characters unhesitatingly accept such beliefs and observe their age-old social customs in their day to day lives. Instead, many of his characters are always in confusion whether or not to accept these. Badang is such a character who wants to rise above such superstition:

> Will I bury my hopes and aspirations because of the fear of such a dream? Should I succumb to such superstitions and go back to the world of grandmother Gidum? No, I can't. I'll take my own route. In any case, a dream is just a dream (*Mon aru Mon*: 134).

The dilemma of the inner self, the conflict within the mind definitely speak of the author's own perpetual doubt regarding following the traditional beliefs of his society. In Badang's thoughts we hear the author's voice on one occasion:

> There was a struggle between faith and disbelief in his mind. He had faced such struggles several times previously (*Mon aru Mon*: 133-4).

Here we see a conflict between indigenous faith and "rational" thinking in Badang's mind. There is a fear of losing one's own identity as a member of an ethnic group. But even otherwise, he is still in danger of losing his identity in the modern rational society if he follows the traditional beliefs and rituals. The important point is that the character is able to overcome the struggle in spite of all the conflict within. Badang, though acutely disturbed and confused, is able to establish his identity in the modern rational world. He does not lose his identity in the face of the modern progressive world. At the same time, he does not shy away from the society where he is born. It is important to note that 'Bodong', a word that closely resembles that of Badang, was the pet name of Lummer Dai and 'Bodong' dedicates *Mon aru Mon* to his grandmother 'Mimi'. So it would probably not be off the mark to say that Badang is a character with which the author identified himself.

Dai portrays grandmother Gidum, the protagonist of the novel, as an epitome of the Adi beliefs, social customs and traditions. On the other hand, the reformist zeal in Dai, the progressive literate youth from the Adi society in him that in later period of his life rose to the post of Director, IPR enables him to depict the character of Badang and Ajong in his novel. Both being literate and progressive minded, they are desirous of change and improvement of the society (*Mon aru Mon*: 116).

Dai appears to be in doubt about the right direction to be taken up by his characters. He shows his characters establishing connections between heavy rain and a person's death as it happens at the time of Gumling's death. On the other hand, according to the popular belief, the sun shines immediately after an accidental death. It is noteworthy that instead of sunshine it rains heavily just after Gumling's death (*Mon aru Mon*: 76). This is a diversion from the popular belief. Dai probably did not want to substantiate the traditional belief that a person's accidental death is immediately followed by sunshine. Dai presents two sets of characters - Gidum and Badang's mother, who unhesitatingly accept the beliefs of their society, represent one group; Badang and Ajong with a scientific bent of mind represent the other. Both of them have the courage to defy the social norms for the sake of the betterment of society and for progress in matters of health, education and way of life. The author's progressive mind cannot allow grandmother Gidum to be cured by *koson* puja but by medicine offered by Dr. Saikia, though he does not dismiss Gidum's deep faith in it.

Stuart Blackburn has rightly observed in his essay 'Tribal Transitions: Cultural Innovations in Arunachal Pradesh, India; Challenging the Romance of Tribal Culture':

Tribal people are not only guardians of culture but also initiators of change (2002: 14).

Both Lummer Dai and his characters definitely fit this description of tribal people, who on the one hand are preliterate people following their traditions and customs; on the other hand, we have the literate people like Badang and Ajong who are always ready to acept change. But it is remarkable that though representing the new and the modern, Badang at one point of time tells Ajong, "I am struggling in the vortex of the new and the old" (*Mon aru Mon*: 118). It is not only the inner conflict of Badang but also the conflict in the writer's mind.

In a biodata circulated by Mrs. Nanni Dai and her family, it is mentioned that "[Lummer] Dai tried to promote and preserve the rich cultural heritage and tradition of tribal society ... One of the strong believers of Donyi-Poloism, Late Dai had founded Donyi-Polo *Gangging* at his residential complex in Naharlagun where Donyi-Polo devotees used to gather every Saturday to offer their prayers" (Biodata circulated by Mrs. Nanni Dai). This proves that Dai is a man with a deep sense of love and regard and attachment to his own culture, tradition and the ethnic beliefs of his society.

To conclude, it would be appropriate to quote Jawaharlal Nehru from the foreword to the first edition of Verrier Elwin's *A Philosophy for NEFA* as it seems that the then prime minister was also in doubt whether to let the preliterate tribal people to live in their own, original set up or to make them aware of the culture and tradition of the outside world:

> It was true that they [the people of Arunachal] could not be left cut off from the world as they were. Political and economic forces impinged upon

> them and it was not possible or desirable to isolate them. Equally undesirable it seemed to me, was to allow these forces to function freely and upset their whole life and culture, which had so much of good in them (Elwin, 1959: Foreword).

Lummer Dai, an upcoming literate youth of then NEFA had also perhaps dwelt in such confusion which made him dangle between tradition and modernity in his novel *Mon Aru Mon*.

In *Kanyar Mulya* (1978), Lummer Dai deals with the social practice of bride price prevalent in Adi society. *Kanyar Mulya* has earned wide acclaim due to its unfamiliar theme that is very much typical of a regional writer like Lummer Dai. The unique cultural systems, traditional beliefs and rituals of the Adi society lend a peculiar flavour to his writings. Homen Borgohain, a prominent writer in Assamese condoled Dai's death by saying

> Lummer Dai is one of the foremost writers of the modern times who has widened the horizon of Assamese literature in the uniqueness of his subject-matters of his writings (*Aaji*, 27th April 2002).

According to the custom of bride price, the marriage of a girl is fixed at an early age, sometimes just after birth, taking the price of the bride from the groom's family. In an interview with Ajanta Dutta, Dai expressed the main inspiration behind his writing of the novel *Kanyar Mulya* that has motivated him to write a novel based on it:

> The case lasted for a long time [in the *kebang*]. It was a case of a girl of nearly 16 years. She was married to an aged man. In the true sense, she was sold to him. In Arunachal Pradesh even now, one has to pay price to marry a girl. Price was paid for that particular girl too. But she was reluctant to be the wife of a person of her father's age. Instead she would love to die, to commit suicide (*Aaji*, Sunday Supplement, 3rd February 2002).

Lummer Dai's words clearly reflect the beliefs of his society. His own beliefs and concept of social reform is also reflected in his answer to a question "Was the ending of the bride really a happy one as you have portrayed?"(ibid. 2002). the *kebang* decided to send the girl to the groom's place by a truck. The ending To such a question he answered: "I do not know. I left the place [Along] when is my imagination only" (ibid. 2002). The interviewer adds that the Adi society that Dai has seen and is acquainted with is reflected in *Kanyar Mulya*. It is not only Gumba of *Kanyar Mulya* who is a commodity but each and every bride of Dai's society is considered so. And perhaps it is because of this that Dai could not bear the sad end of the bride and he has given a new shape to the conclusion (ibid. 2002). It is the sense of duty for his society and a sense of rejection of the age-old custom which makes him see Gumba, the heroine, as free and happy at last:

> Gumba experienced a sweet feeling of freedom; a new feeling of happiness emerged in her mind (*Kanyar Mulya*: 100).

I believe that this rejection of the age-old custom and the happiness of being free from its shackles is a significant departure from the traditional beliefs of the Adis. Though at the outset *Kanyar Mulya* seems to subscribe to the traditional beliefs and social customs, its final message is not one of acceptance and confirmation but rather of questioning and rejection. Viewed in the context of a social practice of bride pride, the novel seems to embody a discourse in which such traditional beliefs are challenged. Such a change is perceived in Borgam Gaken's words:

The rules and rituals of the Adi society has been changed and reshaped (*Kanyar Mulya:* 99).

Kanyar Mulya captures a variety of ideas and themes describing different beliefs and social actions. Greenblatt sees social actions as "embedded in systems of public significations always grasped, even by their makers, in acts of interpretation, while the words that constitute the works of literature are by their very nature the manifest assurance of a similar embeddedness" (Greenblatt, 1980: 5). For him, "pure unfettered subjectivity" is a myth and "the human subject is seen as the ideological product of the relations of power in a particular society" (ibid. 1980: 256). Such an idea is clearly reflected in Clifford Geertz's concept of human beings as "a cultural artefact", and for him culture, far from being acquired or added on to a human beings inherent trait, is "centrally ingredient in the production of that animal itself" (Geertz, 1973: 48).

In *Kanyar Mulya,* voices of different characters in different situations supplement such assumptions and lead us to think that it is not merely the characters of the novel but the novelist himself is also the ideological product of the society. It is this 'ideology' that makes the novelist announce that the characters and stories described in the novel are completely imaginary (In the preface of *Kanyar Mulya*). The contradiction here is his response to a question in an interview:

> I saw the characters of *Kanyar Mulya*. I met them personally. (*Aaji,* Sunday Suppliment 3rd Feb. 2002)

It is ideology that compels Kargum, a major character, to see his daughter as a commodity that can be sold. The same sense of ideology makes the novelist start the novel with the sentence:

> I *sold* the girl a few years back (*Kanyar Mulya*: 1).

And in the next page there is mention of the commodities that Kargum gets as the price of Gumba: "He gets five mithuns, seven cows, eleven dishes and six brass bowls as the price of Gumba" (*Kanyar Mulya*: 2). Kargum is so constructed by society, by the ideology indoctrinated from his childhood, that it prevents him from seeing the girl as an independent human being with her own likes and dislikes. So he says:

> The duty of a girl is to look after the house, to give birth to children, and to look after them (*Kanyar Mulya*: 3).

The author, being the product of such a society, is also not free from such this ideology that enables him to mention the Adi system of giving name. According to the system of the Adi society, the last syllable of the father's name will be the first syllable of his son/daughter's name. The root of such a system can also be traced in the patriarchal structure of the Adi society. The father, being the head, should be remembered everytime one pronounces his/her name.

The novel *Kanyar Mulya* talks about the prevailing social system of bride price of the Adis, the protest raised by Gumba, the new generation, the conflict between the old and the new generations and at the same time the internal conflicts, pains and sufferings of the characters. The society of which the characters are products makes Kargum believe firmly that if Gumba is not sent away to her husband's place, his position as a "respectable" man would be threatened (*Kanyar Mulya*: 14). This society allows him as well as the novelist to believe that Kargum is not doing injustice by selling his daughter in her childhood without her knowledge and therefore, he proclaims:

> I am not doing anything wrong. This is a custom of society (*Kanyar Mulya*: 15).

The social structure of the society allows Gumba's grandmother to believe, "Child marriage is quite a natural and normal thing as many girls are given away in marriage even when they are in their mother's womb" (*Kanyar Mulya*: 17). Ideology compels Gumba's grandmother to consider getting married at an early age, or an arrangement of marriage without the bride's knowledge, or even to sell one girl when she is in the womb of her mother to be a 'natural' and 'normal' thing. In the Adi society slavery was prevalent up to a certain point in time. Just after independence, the government released all the slaves

(*Kanyar Mulya*: 22). Dai introduces the readers to this custom of slavery of the Adi society:

> Talk sensibly. It is because the governement has set you free that you are able to speak here. Otherwise, you still continue to be a slave. Even now people know you as a slave (*Kanyar Mulya*: 22).

Apparently, it seems that the author has tried to portray his society as it is, without any intention of reforming it. But we can say that being indifferent to a social custom of his society, which is the root of so much pain, trauma and agency (not only of a human being but of different people of the society) is neither very easy nor in fact, possible. We can trace the writer's hidden motive to speak for reformation of some age old custom in different situations. In the cultural programme organised to welcome the District Commissioner, the chorus sung by the youths is such an occasion that reflects the writer's motive:

> To take the price of your daughter is to drink your own blood (*Kanyar Mulya*: 30-31).

It is this sense of reform that enables the writer to offer a happy ending to the novel making the heroine Gumba protest against the custom of bride price and breaking free of all shackles of society. Her father Kargum, who managed Gumba's marriage to Mindak's house, wants to give back the price once taken. Lummer Dai's inner self could not ultimately accept the fatal end of a girl due to the age-old custom of bride price. The anxiety within him is whether to follow the tradition and accept the age-old custom or to raise his voice against such a custom which results in pain and suffering and even the end of a life. Therefore, he has argued both for and against the custom of bride price in his novel *Kanyar Mulya*. In the *kebang*, the judicial platform for the Adis, Dai has presented different *borgams* and *gams* championing diverse views. The last chapter of the novel (page 90 to 100) is the expression of the writer's confusion whether to accept or reject the age-old tradition or custom.

Kanyar Mulya is a text that readily lends itself to a feminist interpretation because of the very nature of its theme - the buying and selling of woman, or, in other words, the process of commodification of woman. Kargum is a true product of a patriarchal society. As such, he can announce in front of everyone in the *kebang* that Gumba is the daughter-in-law of Mindak, therefore they can do whatever they want with her. He even grants them the permission of tying her hands if required (*Kanyar Mulya:* 35-36). Though terribly perturbed by his intense love for his daughter and his so-called 'position' in the society, Kargum, the product of the patriarchal society ultimately allows Mindak and his people to snatch her away from her school. His male ego does not permit Minjum to write anything about his tension, anxiety and pain to Mindak while writing a letter to them after that incident (*Kanyar Mulya*: 50 - 51).

There are many characters in the novel who talk of woman as commodity. Jaken, the male chauvinist, once declares looking at Gumba who has been fasting as a protest, "None can say a single word if we bury you alive. Not even your father or you have the right to protest against what we do to you" (*Kanyar Mulya*: 68). Gumba's mother continues to believe that "It is a social rule to keep the girl who does not love her husband in the prison cell of the house"(*Kanyar Mulya*: 71), even when her daughter is in her death-bed after passing several days of inhuman torture, both physical and mental.

The patriarchal note of a male dominated society is seen in its most cruel form in the last chapter of the novel through the voices of different characters. Tumde, an advocate of male domination and patriarchy announces in the *kebang*: "No one has the right to say anything regardless of whether she dies or lives. We have bought her and we can do whatever we want with her" (*Kanyar Mulya*: 91) and "No one needs to argue with a commodity which has been already bought" (*Kanyar Mulya*: 91). At one point Kargum declares: "There is no doubt that she is my daughter, but she is now someone else's daughter-in-law, someone else's wife and so I do not have any right over her" (*Kanyar Mulya*: 93). Similarly, Borgam Itum says: "Even if Gumba prefers to die without having anything in Mindak's house, no one should worry. As Mindak has bought her, he has the absolute right over her. After taking the price for her, Kargum, her father, does not have right to say anything" (*Kanyar Mulya*: 98). Here, Gumba is seen as a commodity - a cow or a mithun - which when bought once, no one else other than the owner has any right over it.

Kanyar Mulya also carries a sense of protest against such male domination. Gumba's protest against the long-established system of child marriage is also a protest against patriarchy. Here, we see the transformation taking place in the traditional Adi society. We hear the sense of protest at the very outset of the novel when Gumba says: "The custom of selling and buying of girls must be eradicated very soon" (*Kanyar Mulya*: 8). Feminism is manifested in the novel in the form of protest against the established patriarchal norms and the attempts to establish identity. Gumba's protest is best reflected when she fasts for sixteen long days in Mindak's house. When the Borgam declares in the *kebang* that "She has decided to die than to marry Dakat" (*Kanyar Mulya*: 90), she refuses and clarifies her position, 'No, I didn't decide that it is better to die than live. I did have desire to live, even now I wish to live. But Mindak's house is like a house of death where I couldn't eat anything" (*Kanyar Mulya*: 90). Another insignificant woman character Yabi's feelings too express a sense of protest against the male dominated society:

> The woman is not a plaything with which you can do whatever you wish. The female body is not a commodity to exhibit (*Kanyar Mulya*: 43).

Gumba, in her death-bed, protests and shouts against

inhumanity, against torture of women, and says. "I am fighting against inhumanity, against brutality and against inhuman exploitation and torture of the women folk by the male of our society. I am fighting against those who have forgotten humanity, morality and I am fighting against a brutal custom of the society" (*Kanyar Mulya*: 65 - 66).

The same sense of change or reform is also heard in many of the progressive educated youths like Minjum. The protest against Gumba's marriage by the students of her school is a new thing altogether in a tribal society of the time in which the novel is set. The chorus and the play that are presented in the cultural function in the presence of the Deputy Commissioner speak against the custom of child-marriage and bride-price in Adi society. The banners and playcards in the hands of the students contain slogans such as - "It is a sin to sell the girl", "Do away with child marriage", "To take price for your daughter is to drink one's own blood", "Behave humanly towards the women folk" etc (*Kanyar Mulya*: 46-47). These slogans and the procession in favour of Gumba's release obviously hints at the transformation of society.

For the understanding of *Kanyar Mulya* (the literary text) one has to look at the local and social context without which any attempt to read the novel may prove meaningless or out of place. The Adi in Lummer Dai can appreciate the fact that Kargum "possesses" rivers and forests:

> How many rivers does he have to eat fish? (*Kanyar Mulya*: 27).
>
> Or
>
> How many forests belong to him? But I do have all (*Kanyar Mulya*: 27).

Similarly, Dai frequently mentions the process of serving *apong* (local wine) to guests.

Upar Mahal was written by Lummer Dai after a long

gap and in it he focusses on the dramatic transformation that has taken place in Arunachalee society since he wrote his last novel. This novel tells the story of how the thirst for money and power converts a naïve tribal man into an inhuman and brutal money-making machine. Dai's previous novel (*Kanyar Mulya*) was published in 1978 and he finished writing *Upar Mahal* in 2002, just one week before his death. The long silence of twenty four years between *Kanyar Mulya* and *Upar Mahal* is too significant to overlook. To seek for the reason behind this long silence and his subsequent re-emergence with a totally different subject matter, we have to look at his biography. Talking about the importance of biography and bibliography for New Historicists, Murfin and Ray quotes Jerome McGann:

> The New Historicist critics will study a literary work's "point of origin" via biography and bibliography. Critics must then consider the expressed intentions of the authors, because, if published, these intentions have modified the developing history of the work (Murfin and Ray, 1998: 241).

Lummer Dai was appointed as the Assistant Director of the Public Relation Department of Arunachal Pradesh in 1979. In 1983 he was promoted to the post of Director of Public Relation department and retired in 1998 as a Director of the same department. After retirement he worked as the editor of the daily newspaper "The Echo of Arunachal" till his death. Kaling Borang talks about the experiences of Lummer Dai while serving as a bureaucrat:

> While in government service and especially in the higher post, he has been associated with people in high places and high society. He has seen ups and downs in the game of power politics and the role of money in the game and other intricacies within. His inquisitiveness and minutely observing author within him couldnot sit idle. He wrote his last novel *Upar Mahal* (*Silifor Review*, 2006: 6).

The society that Dai presents to us in *Upar Mahal* is that of a section of the new generation of Arunachalese who are driven by an insatiable thirst for money and power. This is a society of ruthless ambition, loss of faith, and the politics of power. Being in a high administrative post, Dai had access to the prevailing tendencies in the social, political and economic spheres of life of Itanagar.

The title of the novel *Upar Mahal* means 'higher level' or 'higher department' and obviously it reflects Lummer Dai's desire to depict the life and society, loss of values and inequality, inhumanity and brutality of the people from the "higher level" (*Upar Mahal*: 27). The modern society of Arunachal of which he was a high administrative officer had definitely influenced his views and thinking. Being born in a remote traditional Adi village of Siluk in Pasighat and then holding a high government post in the capital city of Itanagar, Dai had observed the changes taking place in society.

The tribal society of Arunachal has been changing rapidly in the last few decades. The outward change in food habits, dresses and the ways of living has brought about a transformation in the mental set-up of a section of the new generation of the state. No doubt, expansion of education has brought about changes in the outlook of the youth. But at the same time, modernity has brought with it much that is detrimental. Jalam, the male protagonist of the novel, is the embodiment of all the evils of the modern youth - a power hungry up-coming youth who is hypocritic and almost coldly cruel. Extremely fascinated by the idea of being a 'Minister', Jalam is ready to disregard basic human feelings and values. To fulfil his desires he can sleep with any girl (in spite of having a wife and children), that too without any sense of guilt or remorse. He does not shy away from bribing anyone to fulfil his ambition and threatens others without any scrupulousness. Jalam is indeed a product of the new consumerist society that Dai had seen in Itanagar. This helps Dai to make the character of Jalam convincing and realistic.

In *Upar Mahal*, it is the ideology indoctrinated from childhood that makes Jalam proclaim the proud lineage of his powerful clan time and again. He says pompously, "I am the inheritor of a powerful clan" (*Upar Mahal*: 215 - 216). He believes firmly that all his misdeeds would be ignored due to his family name.

The "ideological modes of perception" (Eagleton, 2002:25) enables the writer to look at Jalam, a power hungry young man as caught in the trap of dirty politics of the modern times. The same sense of ideological modes enables Jalam to situate himself in a pre-established 'subject position' - the position of a powerful man from a powerful clan and "reputed" family. The writer's own ideology has constructed him in such a way that he cannot accept the protagonist's remorseless, tension free life as a politician or as a family man. Jalam is corrupt, brutal, inhuman and cruel and he does not hesitate to kill someone who is an obstacle in his way. But the same Jalam is so disturbed at one point that he raises a pistol to kill himself (*Upar Mahal*: 217). It is the ideological modes of perceiving things - due to which the writer is unable to depict a corrupt, guilty person living naturally, normally and freely.

In *Upar Mahal* we see a remarkable departure from the traditional set up of Dai's novels into a completely different setting. Lummer Dai has written four novels where he talks of one or the other Adi rituals, tradition or customs. In *Upar Mahal* he talks of a newly formed Arunachalee society in general. This Arunachalee society is the composition of many tribes as well as members of non-tribal community. The last years of his life in the capital complex, in a society with people from various tribes as well as from various states of the country has definitely shaped his views differently.

In this novel Dai talks about different groups of people. He depicts the Adis, Monpas, Khamtis, Noctes etc as well as the Assamese, Bengalis, Biharis and communities from South India. The population in the capital complex is increasing every day because of the migration of these people. The writer has beautifully portrayed these people occupying various posts and professions. But the significant point is that the protagonist of his novel *Upar Mahal* too, is an Adi. As a conscious Adi intellectual, Dai probably felt the need for reform within his own tribe before pointing a finger at others. Arunachal Pradesh is the homeland of more than twenty six "major" tribes. It is perhaps due to this diversity in their languages, customs or food habits that almost all the different tribes have become even more conscious of their own identities leading to the formation of representative socio-cultural organizations - the Nishyi Elite Society or the Apatani Socio-Cultural Society etc. It is important that each member of such society is so contented and limited to his/her own society that he/she develops a kind of indifference, and sometimes even hostility towards the members of other societies.

In *Upar Mahal,* we see the politics of inter-tribal rivalry quite clearly; members of the same tribe very often use one another to their mutual advantage. Jalam (the Adi Home Minister) transfers the police inspector Wangham (a Nocte) and inducts the newly promoted Tapam (the Adi) in his place. We see the association of Tapam with Jalam as they are both from the same tribe. This association between Tapam and Jalam is the writer's direct expression of the present Arunachalee society; there is a greater level of trust between members of the same tribe than otherwise.

The complex social and political forces of the times in which the writer lives and writes has shaped the novel. There might not be any specific, historical incident on which the theme of novel is based. Stephen J. Greenblatt rightly says in his essay "Resonance and Wonder":

> New Historicist critics have tried to understand the intersecting circumstances not as a stable, prefabricated background against which the literary texts can be placed, but as a dense network of evolving and often contradictory social forces. The idea is ... to situate the work in relation to other

> representational practices operative in the culture at a given moment in both its history and our own (Greenblat, 1990: 170).

According to the New Historicists, the text is never isolated from the context. Nanni Dai, the author's wife, was afraid that publication of the novel (even as Dai was in the process of writing *Upar* Mahal) might jeopardize her husband's life (Nath, *Dainik Asom*, March 22 1999). This proves that *Upar Mahal* is indeed as close to the present Arunachalee society as possible. Here we see that "far from being divorced from the world, literary production is itself a form of social practice: texts do not merely reflect social reality, but create it" (Patterson, 1995: 260).

Foucault saw history in terms of power; he viewed power as a complex of forces that produces what happens. It is this power "as a complex of forces" that produces the complex actions in *Upar Mahal*. It is the power that enables Jalam to act as a fearless, upcoming youth of the new generation who has the license to behave in the way he wants. He is so constructed by beliefs and practices that he can behave the way he does. It is this "power" that makes him declare "I am not the one to bow down in front of anyone, even if people call me corrupt, immoral and murderer" (*Upar Mahal*: 215).

At the same time, Jalam is also empowered by discourses and practices that constitute power. Therefore, in spite of being involved in crimes like bribing, corruption, rape and murder he can deliver sermonizing political lectures saying, "Our government should be transparent... we should be accountable, be faithful to the general people. We must maintain law and order strictly... you can even arrest me if I am accused of being involved in any crime" (*Upar Mahal*: 134).

Jalam is "powerful" enough to have illicit relationships with Yabin, the dumb labourer and Yaka, the poor girl who submits herself to him just for a job. In neither case is Jalam repentant as he has the "power" to use both of them only as things that he can use for the satisfaction of his physical desire and can do away with later on, because both the girls are from relatively poor families. Jalam appears to be a bit disturbed because of his relationship with Doima who is ready to give birth to Jalam's child. In this regard, it is important to note that Doima is different from Yaka and Yabin. Doima is a talented, beautiful, highly educated girl whom even Jalam's wife is ready to accept as his wife, "I welcome her, Drema can stay with me."(*Upar Mahal*: 195). The same Tempi (Jalam's wife) is not ready to accept Yabin (the dumb labourer) even though she is sure of the fact that Jalam is the father of the child as the boy looks exactly like Jalam. The significant point to note is that Tempi is ready to accept the boy as their own son:

> The boy is your's. He looks exactly like you - and they have informed me that he is your's. And now I accept that he is ours (*Upar Mahal*: 187).

Tempi's acceptance of the boy as their own son is also a reflection of an age old belief in the Adi society. As in all patriarchal societies, the Adis consider the boys as the inheritor as well as a means of reproduction of the family; therefore, a boy child is always welcome in a family. The same idea is expressed in Jalam's words when he is disturbed by Drema's pregnancy:

> What if Drema gives birth to a boy? Boys are always necessary to strengthen the clan (*Upar Mahal*: 180).

A parallel theme of patriarchy is also being observed in the novel. When Yabin feels happy to have a child that resembles Jalam, we are reminded of an Adi tradition: "Yabin might have forgotten that the father has absolute right over the son" (*Upar Mahal*: 168). Jalam's wife, Tempi too raises her voice against the patriarchal social structure (*Upar Mahal*: 186).

5

Ethnic Worlds in Yeshe Dorjee Thongchi

Yeshe Dorjee Thongchi, a recipient of the Sahitya Akademi award in 2005, published his first novel *Sonam* in 1981. Thongchi's novels generally deal with the traditional, cultural life of the Monpa and the Sherdukpen tribes of Arunachal Pradesh. *Sonam* is no exception in this regard. The novel is centered on the Brokpa tribe (a section of Monpa), a community owning the Yak herds. The family system of the Brokpas, where a woman is socially allowed to marry more than one man is the custom in which the story of the novel is rooted. Such a unique system is called *khor depka* by the Monpas. The setting of the novel against the backdrop of the Dirang and Tawang districts of Arunachal necessitates the mention of the life-style, customs and culture of the society where the domestic yak-oriented economy is the mainstay of the locals.

In a given historical moment, different modes of discourse such as law, theology, morality, philosophy, literature, art, architecture etc. are not autonomous, but the constitution of a given cultural field and of some ideological codes that order all discourse in that particular culture. Therefore, a literary text is always subject to the particular conditions of a time and place. Thongchi's *Sonam* is also subject to the particular conditions of a time and a place - the time being the middle half of the 20th century and the place is that of the inhabitance of the Brokpas (the Dirang and Tawang Districts of Arunachal Pradesh). Thongchi himself has said that he has tried to draw the picture of the Brokpa tribe and the polyandry system prevalent in the society (in his preface to the revised edition in 2005).

A literary text needs to be placed within the 'frame' of a non-literary text in the sense that it can very often be seen to bring together the theme or subject-matter with one or some historical document. Apparently, *Sonam* deals with no historical happening or document but reflects the culture and society of the Brokpa tribe of Arunachal of a particular time. Having stayed among these people at Tawang the novelist has seen, met and interacted with them from very close quarters and thus *Sonam* is not only a single literary text with a triangular love story but a social document of the Brokpas.

In the novel *Sonam*, Thongchi has not used any particular context but the entire novel is a co-text of the traditional beliefs, rituals and culture of the Brokpa tribe and the story of the love and hatred, the ecstasy and trauma of Lobjang, Sonam, Pema Wangchu and Tsering Drema. A literary text is an interactive component within the network of institutions, beliefs and cultural power-relations - practices and products which constitute what we call history. The novel *Sonam* can also be read as a convergence of different beliefs, power-relations, cultural practices, and ideologies of a particular society. *Sonam* begins with the description of a robust young shepherd of the Brokpa community coming back home from the Choksum hills to the Sakteng village.

With the progress of the story, Thongchi introduces us to a variety of social customs and traditional beliefs as well as the lexical and cultural world of the Brokpas. The livelihood of Lobjang (the main character) of rearing Yaks and going to far off places like Tawang, Bhutan,

Kalaktang, Rupa, Dirang to sell or barter *ghee*, *mar* (milk fat), *churpi* (a type of cheese made from milk), and sheep-wool gives the idea of an actual Brokpa man with an animal farm away from home. At the very outset of the novel we are introduced to a traditional belief when Lobjang thinks of bringing his wife Sonam to his farm and staying there together:

> If Sonam was taken to the farm and they start living as husband and wife the sanctity of the place would be violated. Any physical intimacy with the wife would incur the displeasure of the Sungmu, the household deity of wealth, whereby good fortune would forsake the place and epidemics would break lose upon the animals of the farm (*Sonam*: 15).

Lobjang does see his and his wife's social roles as inevitable and natural and he does not envisage the possibility of any change.

It is due to the 'power', the power that the men of the society like Lobjang enjoys, that we see him acting according to his whims. He blames Sonam for all the problems in his life and even when he is invited to have a meal with her, he refuses. It is this power that Lobjang uses to exercise his will in the lives of his wife and their children. It is he who makes Pema Wangchu a *khor depka* for Sonam and takes both his children to the Tawang gompa to make them *anei* and *drapa*. However, the fact that the same sense of power that Lobjang enjoys can also ensnare him in its own trap is proved when he is unable to go to meet Sonam, his beloved wife, even when she is acutely ill.

Sonam is basically about the sorrow and agony as well as the love and hatred among the three main characters caught in a love triangle. Apparently it seems that the women enjoy more freedom in the society that the novelist presents in *Sonam*, but while going through the novel, one discovers that it need not necessarily be so. The community is not only male-dominated but abounds in ideologies and beliefs that help in gender construction. Polyandry and polygamy are permissible in Lobjang's community and we hear him talking to his wife to get another groom for her in a matter-of- fact manner. When he discovers Pema Wangchu in his house with his wife, he says, "If your mind is set on Pema Wangchu, let us bring him into our household" (*Sonam*: 21) and "If you are not interested in Pema Wangchu let me look for another boy. Tell me is there anybody else that you would like to be a co-husband to you?"(*Sonam*: 22).

It is clear that Lobjang is not at all worried about sharing his wife with someone else. But at the same time, the intention behind such a proposal is questionable. Polyandry is prevalent in the society of the Brokpas to which he belongs and he is so constructed by the ideologies and beliefs of his society that even the idea of sharing the most cherished woman of his life does not create any trauma or agony in his mind. Instead he is happy to get a man who would help him in his household activities. Lobjang, being the male 'subject' of the society has the 'power' to decide and Sonam has to accept it in spite of her doubts and uncertainties.

In *Sonam*, the strong influence of patriarchal ideology is manifested at different situations in the opinions of the characters. The strong note of patriarchy prevalent in the society is best reflected in Lobjang's faith in an age old belief of violating the sanctity of the farm by Sonam's coming there and staying together with her husband. In Lobjang's words we hear the bias of a typical patriarchal man when he is alone and is on the way to Pema Wangchu's house:

> Women are like dogs, whoever pats them on their back-they will meekly follow him and will snarl and snap at him who once in his life speaks or behaves badly. (*Sonam*: 86)

The most significant aspect that deserves to be

mentioned about the novel is the writer's addition and extension of some ideas and chapters in the new edition of the novel in 2005. Thongchi has tried to "justify" the changes:

> At the time of writing the novel, I did not have much knowledge about the traditions and rituals, customs and culture of that society. Polyandry is a practice of the Brokpas (the yak owning community) and such a practice of marrying more than one man is called *khor depka* by the Monpas. Due to my ignorance the words like Brokpa, *khor depka* etc. were not used in the first edition of the novel (Preface to *Sonam,* 2005).

But the changes in the new edition of the novel go far beyond the inclusion of words from the Monpa society. A New Historicist reading of the novel forces us to concentrate on the significant changes brought about by Thongchi in the new edition. We also need to look at the biographical details of Yeshe Dorjee Thongchi, a top-level bureaucrat, an intellectual, and a progressive member of Arunachalee society while focusing on the new ending of the novel.

The first edition of *Sonam* concludes with Lobjang's departure to the Tawang *gompa* along with Rinchin Jangmu and Tashi. But the progressive man in Thongchi is no longer happy with the same conclusion. He feels the need to speak more about *khor depka* - about the age-old tradition of the Brokpas whereby a woman is allowed to marry more than one man. We can hear him expressing his own belief towards the end of the novel through the mouth of Tsering Doima:

> My sister warned me not to repeat the mistake made by her. She advised me not to get involved with another man to encourage *khor depka* (*Sonam*: 128).

And

> We won't live as Brokpas anymore. We will go to Tawang and buy some land to cultivate. If we remain Brokpas, I will not always be able to accompany you. I will be compelled to perform *khor depka* again with someone else (*Sonam*: 128).

Being born to the Sherdukpen society which is very close to the Monpa society, Thongchi has observed the Brokpas from close quarters. In the first edition of the novel, he was happy to portray life as it is. But in the new edition he felt the need for its extension - showing Lobjang's repentance of his action and his willingness to act according to the wish of his dead wife Sonam. One discovers a new urge in Thongchi to reform society.

Thongchi's change in mind therefore has a strong connection with the change in his public image. Greenblatt notes that Shakespeare's young son Hamnet had died shortly before his father began work on the play (Gewertz, 2004). Shakespeare would have been in Stratford for the boy's funeral, and it might have been possible that Shakespeare's Catholic father urged his son to have prayers said to speed the child's release from purgatory. Shakespeare must have still been brooding in late 1600 and early 1601, when he sat down to write a tragedy whose doomed hero bore the name of his dead son. His thoughts may have been intensified by news that his elderly father was seriously ill back home in Stratford, for the thought of his father's death is deeply woven into the play. And the death of his son and the impending death of his father—a crisis of mourning and memory—could have caused a psychic disturbance that helps to explain the explosive power and inwardness of *Hamlet* (Greenblatt, 2004). The problem was that purgatory had been abolished by the ruling Protestants, and saying prayers for the dead declared illegal. Hence, the possible dilemma for Shakespeare, and for many English people who still felt strong ties to Rome, was whether to risk punishment by praying for their deceased loved ones or obey the law and allow those souls to languish in flames. This anxiety regarding one's obligations to the dead, Greenblatt suggests, lies behind Hamlet's indecision about whether to obey his father's ghost and take revenge on his uncle Claudius.

Lingjik (1982, 2003), the second published novel of Yeshe Dorjee Thongchi is mainly about the traditional marriage system of the Sherdukpen tribe. The novel expresses the clash between two generations - the old representing the characters like Penje, Ajom (Premtasi's uncle and aunt) and Lam Doima and Kejang (Premtasi's parents) and the new generation represented by Premtasi. The conflict (both physical and mental) in the characters as well as the progress of the story is rooted in this clash of two generations. Jeetanjali Borpujari has rightly observed:

> A realistic picture of the clash between the new and the old is depicted vividly in the novel *Lingjik* which is written in the backdrop of the Sherdukpen society (2000: 119).

While we read between the lines of the novel, the writer seems to be very sensitive to the issues concerned. He is trying to observe his society (the then Sherdukpen society) from the eyes of a reformist. His reformist ideas and views are summed up in the character of Premtasi, the male protagonist. Dilip Bora has commented:

> The dark sides of the tradition and customs of the tribal society as well as the best things in it are brought into light in this novel. The author is depicting the rules and rituals, tradition and customs of the society with great honour at the same time expresses a wish to change these (Bora, 2000:137).

In the light of this statement, we can analyse the novel *Lingjik* and its author. Yeshe Dorjee Thongchi, a Sherdukpen was a student of Cotton College, Guwahati while writing this novel (as stated in the preface). He is perhaps the first generation member of his tribal society to leave his state to go to Guwahati for higher education. Moving out of his society which was very close, self-dependent and self-contented, Thongchi saw a new world, that was wide and free, and so, different from his own. He was dwelling between the attachment and loyalty to his root, his tribal background as well as his desire for a better, sophisticated and modern society. This duality within allows him to write the novel *Lingjik*. Therefore, we observe a sense of respect and love towards the age-old customs, beliefs and tradition along with the urge to reform many aspects of tradition.

The first sentence of the novel and the repetition of the idea of stumbling in it offer a clue to the understanding of it. The novel begins with a sentence "Premtasi has stumbled" (*Lingjik*: 1). The sixth chapter begins with a sentence "Penje has stumbled" (*Lingjik*: 36) and the eight chapter begins with "Dorjee Tsewang has stumbled" (*Lingjik*: 53). The word 'stumble' is used metaphorically by the author to express a sense of shock that is sudden and unexpected. Premtasi puts a step forward towards a world of new values and beliefs and in this he stumbles. Stumbling is not only about physical stumbling but also about mental staggering. The author talks about the psychological stumble through Penje's word:

> This is a simple stumble. But the bigger one is in my life; the wound of that 'stumble' will never be healed (*Lingjik*: 43)

Through these words of Penje, the author has made the readers conscious of the metaphorical meaning of the word. In the first chapter of the novel, it is Premtasi who stumbles on a stone and wants to break it into pieces, but the subject-position granted to him by ideology (as an EAC) prevents him from taking out his anger on a lifeless object. Then, it is his uncle Penje (father of Dorjee Tsewang) who stumbles and severely injures himself and then breaks the stone into pieces. He takes revenge upon the stone with all his strength. Next it is Dorjee Tsewang who stumbles on a stone on the way back home and hurts himself so seriously that he loses consciousness for three days. The repetition of this action of 'stumble' is also definitely the author's conscious attempt to offer a kind of importance to a stone - a lifeless object. Stone is a very important symbol in the novel. We infer that the author is doing this just to make the readers conscious of the importance of a stone pillar, 'lingjik' which is not only a pillar of stone, but an embodiment of deep faith in social customs and traditional beliefs.

We can see striking parallels between the author and the male protagonist Premtasi. Premtasi is an Extra Assistant Commissioner (*Lingjik*: 2) who 'stumbles' against a stone on the way back home from Mr Tyagi's house. His immediate reaction to this is to break the 'stone' into pieces just as revenge. But he realizes "It does not suit a person like him to behave so" (*Lingjik*: 1). "A person like him" strengthens the difference between Premtasi and the others as he sees it. The author sees him as someone who is an EAC and so does not or cannot behave like a "common" person. Yeshe Dorjee Thongchi, while writing this novel was a student who might have seen his future in the character of Premtasi. Incidentally, Thongchi later became a District Commissioner who served in different districts of Arunachal Pradesh and is at present serving as the Commissioner to Government of Arunachal Pradesh. Being an educated youth and government officer, Premtasi finds it difficult to accept the forceful capture of his educated sister Krima Lamu by his cousin Dorjee Tsewang according to the tradition of the Sherdukpen society. Therefore, he breaks the marriage of his sister Krima blaming Dorjee Tsewang for his impotence. But this is what gives rise to a feeling of animosity and leads to a conflict between the families of Premtasi and Dorjee Tsewang. Feeling very insulted Dorjee Tsewang asks for the bride price of Lam Doima (his aunt) from Kejang. Asking for bride price in the life time of the victim is believed to be very shameful and insulting in the

Sherdukpen society. This causes mental torture, sorrow and pain in a few characters. Towards the end of the novel, Premtasi finally realizes that age old customs and traditions cannot be easily abolished, and that in fact, it might not be really essential. Changes cannot be brought overnight in societies.

Being insulted by his collegues and friends Premtasi tries to change the social norms, - a particular system in case of his sister's marriage. This creates trouble, not only in his mind but in the life of many characters. In Premtasi's decision to break his sister's traditional marriage, we hear a sense of protest against the age-old customs, the old world. At the same time, his realization of the fact at the later part of his life that such a change would definitely create problems, both internal and external, places Premtasi in the position of a person who has deep faith and honour in the age-old customs, traditions and beliefs of his society. We can compare the writer's own dual self with the character of Premtasi.

Lingjik is also about a longstanding custom of the Sherdukpen community. 'Lingjik' is a pillar of stone which is set as a token of a very big decision, as the boundary line, to remind the heirs of the clan about that particular issue. The title of the novel speaks of a social custom, and at the same time there is mention of various traditional beliefs. In the words of Premtasi, we are introduced to the traditional Shurdukpen marriage system (*Lingjik*: 3), the concept of 'samu' - the niece who goes to the maternal uncle's house as a daughter-in-law and even about the folktale associated with the Bobo Thekong Kong bird (*Lingjik*: 26). Moreover, there are mentions of *khalpu* (the middle man) who takes part in negotiating between two people during a misunderstanding or battle, of the game of pulling hands to test the strength among the young men and so on. This is the reflection of the writer's attachment to the society, of the social and ideological world he is living in.

Premtasi is not a man of unflinching faith in each and

every social custom of his society. He is confused whether to accept the age-old customs of his society blindly or to reform it for the betterment of the society. At one point of time, Dorjee Thongchi, perhaps suffered this confusion as an upcoming educated youth of his society. So, he expresses his views in the words of Premtasi, who after being laughed at by his non-tribal colleagues and friends about the traditional marriage system, exclaims in anger:

> Shut up! What do you think of me? You have come here to serve our people, not to laugh at our traditions. Be careful! (*Lingjik* : 4).

Premtasi wants to be a part of the modern society and he believes that it is possible only after doing away with the old customs and superstitions. Both Yeshe Dorjee Thongchi and Premtasi were exposed to other cultural values through modern education and socialization and so they tend to identify some customs of their society as superstitious. They are both brought up in accultured situations.

The anxiety of the author's dual self is summed up in the character of Krima Lamu. Once she is presented as a traditional, simple, docile, obedient and shy girl who does not have a say in the decision of her marriage, who cannot even think of disobeying or protesting against elders:

> She felt as if her hands and legs are tightly tied with a chain and she had no option but to listen to what was talked about (*Lingjik*: 21).

But the same girl is throbbing with courage and strength and scolds her brother Premtasi, her uncle Penje and her father Kejang in another situation. She protests against male domination in a society where women are believed to be a commodity and asks the men to give her back her sense of identity. Her remark (*Lingjik*: 117) definitely conveys the image of a vey strong and bold girl. This shows the author flirting with contradictory ideas regarding a girl's strength –as simple, traditional and obedient and also as conscious of her identity, ready to protest for the injustice committed against her.

We see Premtasi bringing back his sister Krima and then educating her. Finally, Krima becomes a doctor. But the journey for Krima from her in-law's to the position as a doctor is not smooth and simple at all. By bringing back Krima from her in-laws, Premtasi not only hurts the Dorjees (the groom's family), but also hurts the age-old belief of the people of his society. Therefore, Dorjee Tsewang (the proposed groom for Krima) asks for bride price from Kejang for his aunt Lam Doima, even though taking bride price in the life time of the victim is believed to be very shameful and insulting. Moreover, a 'lingjik' is also put up to remind the coming generations of an everlasting battle between the Wangja lineage (Dorjees) and the Thongo lineage (Pemtasi). According to the traditional belief, after the 'lingjik' is put, there cannot be any mutual understanding between the two lineages even in future. As such, this causes anxiety not only to individuals but also for the society in general. Lam Doima dies out of frustration and humiliation. The same Premtasi who is once ready to do everything to bring his sister back home, seems to be scared when Dorjee Tsewang puts up the 'lingjik'':

> He is scared at first when he sees that Dorjee Tsewang is going to break the relation between the Wangja and Thongo lineages ... people laughed when he requested them not to do so. He was not even allowed to leave the place at that moment. As such, he had to witness such a historical decision (*Lingjik*: 98).

That Premtasi's confidence is jolted by the setting of a stone-pillar is an important matter. We can conclude that the author, however free and rational in his thinking, is perhaps unable to detach himself from the society and therefore, cannot depict Premtasi as remaining unaffected by ideologies.

We see Prematasi in the first chapter as an educated upcoming youth who wants to protest against the age-old customs of his society. But towards the end of the novel, the author gives a different picture of Premtasi, one who can bow down in front of his people to convince them about something (*Lingjik* : 96).

Premtasi does not like the traditional marriage system of abducting the bride. But the same man is disturbed by the custom of setting of the 'lingjik'. This obviously says something about the author's own self. Being educated and influenced by the outer world, Dorjee Thongchi is perhaps more conscious of the different dimensions of the age-old beliefs and superstitions of his society. At the same time, he is aware of the traumatic consequences if such customs and beliefs are checked all of a sudden. As the people, in each and every social group is so constructed by the ideologies of that society that it prevents a person from objetively seeing his/her own position in it.

Ideology helps one to place oneself in an imaginary subject-position and a removal from that subject-position might lead to severe mental anxiety to that person. Premtasi's mother Lam Doima is so constructed by ideology that she cannot bear taking bride-price from her husband even when she is alive as it is considered to be very shameful. Ideology does not allow her to remain unaffected by such "a shameful act" and it finally causes her death. We can say that the author Thongchi is also not free from the ideology that he has indoctrinated from his childhood due to which he is unable to depict Lam Doima as surviving the ordeal. The author, so constructed by ideology, cannot see her rise above such social constructs. Instead he lets her succumb to the age-old custom which ultimately leads to her death.

The novelist tries to show a kind of transition in his society through the character of Premtasi. But he is not sure of the way of doing that. Therefore, in spite of being a highly educated government officer, Premtasi is in doubt after the breaking of the engagement of his sister, Krima and Dorjee Tsewang. On one occasion he declares:

> I want to change the marriage-system of our society lest our next generation spoils anyone's life by committing the same mistake (*Lingjik*: 29).

But the same person feels 'guilty' (*Lingjik*: 29) after the departure of his uncle-aunt (Dorjee Tsewang's parents). He cannot sleep at night; he is disturbed by many thoughts. He is in perpetual doubt about the necessity of his action, "He could not decide whether he has done the right thing or not" (*Lingjik*: 28). The novelist's own self is reflected in Premtasi's character. Being an educated person, he wants to wash off the superstitions and bad practices of his society which he sees as an obstacle in the progress of the society. But he does not find any new way of abolishing the old.

Premtasi's sister Pema is illiterate and she does a lot for him, including managing his expenditure in the college and frees him from the responsibilities of the house-hold works as the eldest child of his parents. So he is very grateful to Pema:

> He is very grateful to Pema who had helped him by arranging his expenses during the college-days. In spite of all these he was unable to buy her a new dress (*Lingjik*: 35).

Yeshe Dorjee Thongchi has dedicated this novel to his elder sister, Thar Doima who is illiterate and who helped him a lot in writing the novel by "lighting the lamp" and "offering tea in the writing table" as he tells us in the preface to the novel. There is an obvious parallel between the character of Premtasi and Thongchi. Another important issue that proves this parallel is the mention of Premtasi's would-be bride by his uncle and aunt, Penje and Ajom. Penje says, "He won't agree to marry anyone whom his parents choose. He wants a 'memsahib'" (*Lingjik*: 43) and Ajom supplements him by saying, "He is already attached to someone from a different community ... He is going to ruin his parents' reputation" (*Lingjik*: 43).

His mother Lam Doima talks to Premtasi angrily about the *kakpu* (lower caste) girl Chebom whom he loves (*Lingjik*: 78). Chebom is a Monpa girl (*Lingjik*: 104). Thongchi himself married a girl from the Nyishi tribe. Thongchi married a girl from another tribe for which he might have a 'guilt-feeling' which does not allow him to show the marriage of Premtasi and Chebom at the end of the novel. Premtasi's mother curses him: "The sin of breaking up the marriage of Krima and Dorjee Tsewang will harm you" (*Lingjik*: 79). It is this sense of 'sin' or 'guilt' that prevents the novelist from depicting the reunion of Premtasi and Chebom.

Towards the end of the novel, Premtasi accepts his guilt, the guilt he has committed by breaking the marriage between Krima and Dorjee Tsewang:

> I am playing with your life; with the life of my mother ... I am guilty. Education has not opened my eyes, instead it has made me blind. I have started to look at my own society as being inferior and I think of myself as being different from others. I am pushed into darkness by the outward glitter ... (*Lingjik*: 117)

It is the realization of his idealism inspired by which he wants to change the foundation of his society. We hear the novelist's voice in Premtasi's thoughts:

> He knew that he was wrong at that time. Then he was a romantic idealist and with that idealism he wanted to change the society. He didn't know that a society can't be changed in one day. He didn't know that it required vast knowledge, deep experience and patience (*Lingjik*: 132).

The novelist's faith in his root, in his traditional social life ultimately brings Dorjee Tsewang and Krima Lamu together. Though they suffer a lot in their lives, Krima and Dorjee Tsewang start a new life together. And it is this faith in the root that does not allow Premtasi to be happy with Chebom, a Monpa girl. The novelist's faith in his own beliefs and tradition, culture and custom prevents him from showing a union of Premtasi with Chebom and instead shows him getting married to Langa.

Before the reunion of Krima and Dorjee, the novelist feels the urge to remove the 'lingjik'. Both Dorjee and Premtasi manage to break the 'lingjik' - the witness of trauma and agony of many persons. This act of breaking the 'lingjik' is a step towards the progress of the society and peace of the people.

Mouna Ounth Mukhar Hriday (2001) is Thongchi's third published novel. The English translation of the novel was released in 2010. We find a kind of juxtaposition of the universal and the particular in this novel. While dealing with the universal theme of love, the author also talks about various traditional rituals and customs, beliefs and cultures of his tribal society, i.e., the tribal society of Arunachal Pradesh. Language, used by the author in this novel is a significant point that deserves mention. The author has made it clear in the preface that "due to the necessity of the setting and subject-matter of the novel, the same character speaks in different languages" (Preface to *Mouna Ounth Mukhar Hriday*). We observe that when a non-Assamese or tribal character talks to people from other tribal group, then he/she uses broken Assamese, and they are shown to speak in Assamese while talking within the same tribal group. Though the writer has tried to substantiate his point, we can explain his deliberate use of a variety of languages from a different perspective. The novel is set in the Arunachal Pradesh of the nineteen fifties when the state (then NEFA) was quite cut-off from the rest of the world. Communication was very difficult and so developmental activities took a lot of time. The novel tells us about the construction of a road from Saku to Bomdila which is a real incident. At the same time, it is also mentioned that for the construction of the road, youths from different tribes of different places of Arunachal Pradesh were gathered. The significant point is that though Arunachal is a land inhabited by many tribes, the language of one tribe is not known to a member of another tribe. Hence, Assamese plays the role of a "lingua franca" for them.

Dorjee Thongchi, being a Sherdukpen from a village just thirty kilometres away from Bomdila, might have seen and experienced the difficulties in the road-construction site. He had probably undergone the difficulties associated with it and had communicated with the people himself. The broken Assamese used by the non-Assamese and the member of one tribe while talking to a member of another tribe is a realistic scene of that time as the people at that point in time were more accustomed to the Assamese people and Assamese language. At the same time, due to the geographical barriers and lack of communication, one tribe from from East Siang is like someone from the outer world for some other tribe from the West Kameng district in the world of the novel. And it is perhaps the sense of surprise, fear and hatred on the part of one tribe toward another tribe of the same state that has inspired the author to depict a relationship of intensely passionate love between a girl and a boy from two separate tribes.

The love affair between Rinchin and Yama (a Sherdukpen boy and a Nyishi girl) forces us to look at the author's own biography. Yeshe Dorjee Thongchi, a Sherdukpen, married a girl from the Nyishi tribe. On the other hand, he has dedicated this novel to his wife Sona. We find a parallel between the protagonist's and the writer's lives. However, the significant thing to note is the inability on the part of the writer to see Rinchin and Yama happily united at the end. In spite of the writer's intense urge for bringing about unity among the various tribal people of Arunachal Pradesh, he is unable to let Rinchin marry Yama. The tragic end of the novel is perhaps a result of the writer's own guilt-feeling of having married someone from another tribe.

In the novel, Thongchi's characters, including Rinchin Norbu, consider it to be a "matter of shame" to be romantically involved with a member of another tribe and a "sin" (*Silent Lips Murmuring Hearts*: 44).

Mouna Ounth Mukhar Hriday marks a departure from the conventional subject-matters of Thongchi. While *Sonam* and *Lingjik* are based on a particular social custom of the Monpa and the Sherdukpen tribes, *Mouna Ounth Mukhar Hriday* is not rooted in any particular social custom. Instead it is based on the universal theme of love between a boy and girl who speak different languages but share the same language of the heart. But just as every person is a product of his society, of the culture he is born and brought up in, so is the author.

This novel leads us to an understanding of the culture in which Thongchi is born. We are introduced to the variety of rituals, customs and traditional beliefs of the various tribal groups of Arunachal Pradesh through this novel. In his first two novels Thongchi restricts himself to his own tribal society, i.e., the Sherdukpen and the Monpa societies with which he feels a kind of affiliation.

Incidentally the first two novels of Thongchi were written during his college days. But *Mouna Ounth Mukhar Hriday* was written after a long gap in 1991 when Thongchi was serving in the Arunachal Pradesh Public Service Commission. This conversion of his status from a student to an Administrative Officer had definitely altered his views. In his third novel he tries to bring together the people of his state in a bond of unity. Being an administrative officer in Tawang, Lower Subansiri, and Chanlang districts he must have witnessed the sense of fear or hatred, a sense of indifference in the tribes towards others. He dreams of the development of the state with educational institutions, hospitals and industries that would crop up and that these would be managed and run by the people from the state:

> A day would come when this land of NEFA could be considered a developed place. There would be schools, colleges, hospitals, and industries in these hills and the local people would run these as teachers, doctors, engineers and industrialists. The day when the light of education would illuminate

> these hills was not very far off (*Silent Lips Murmuring Hearts*: 132).

The author is hopeful not only of progress but also of a feeling of unity and brotherhood among various tribes. Thongchi depicts the self-contented lives of the tribes and shows that they hardly had any interaction with members of other tribes. Each of the tribes considers themselves to be the best in terms of caste, beliefs, dresses and rituals. We hear Yado saying:

> Today a new group of labourers has reached from somewhere. Strange clothes and strange language! Before coming here I never imagined that there were so many races of people in this world of Donyi Polo, besides the Nyishis. I thought the world was occupied only by Nyishis and Sulungs (*Silent Lips Murmuring Hearts:* 23)

Yado, the Nyishi girl is so constructed by the society she is born and lives in that the people with different dresses and languages seem strange and alien to her.

The author is conscious of the diversity of language, dress, rituals, customs and traditions of different tribes of the state. At the same time, he is conscious of the sense of pride and superiority of tribes compared to others. It is perhaps due to this consciousness, that the characters of the novel are portrayed either as being completely ignorant of the other tribe or as having a sense of racial superiority. We hear this sense of superiority in Tadak, a Nyishi youth's words:

> The Ngisaks are a lower race than us. Amongst all the races in this world, our Nyishis are the highest. We collect tax from the Bangros, and the Bangros collect tax from the Ngisaks. The plight of the Ngisaks is comparable to the plight of a mouse that suddenly comes across a cat when coming face to face with the Bangros – the very Bangros who are so scared of us (*Silent Lips Murmuring Hearts:* 24).

On the other hand, the same sense of superiority is reflected in the words of Tasi, a Sherdukpen girl:

> All the same, I am terrified of them [the Nyishis]. You don't know where they are from, what race they belong to, what they eat or what they wear. How can you trust people from other tribes? It would have been better if we were allowed to work with our own people (*Silent Lips Murmuring Hearts*: 31).

The Sherdukpen youths are surprised to see the strange, unknown faces of the Nyishis in their new dresses and customs and the same was the case with the Nyishis when they set eyes on the Sherdukpens. Having worked in different places of Arunachal Pradesh, Thongchi has perhaps some personal experience of such views among different tribal people. Therefore, he is capable of drawing these characters so vividly and naturally. But, the intention behind this can be summed up in his attempt to bring them together. He has shown the development of the love-affair between Yama and Rinchin who belong to different tribes. At the same time, Thongchi shows his characters finally coming to the realization that all people are the same with the same human values, love and friendship:

> We Bhutias think that we are the greatest race while you Daflas think that you are the greatest. That is why we hate each other. We consider each other our enemies and fight. But having stayed with you during the construction of the road I have come to know that we are the same. It's just that we speak different languages, follow different customs, and wear different clothes but our minds are the same. (*Silent Lips Murmuring Hearts*: 123)

Here, we sense the novelist's urge to reform society, the urge to pass a message to the people of his state. In different situations of the novel, the writer has expressed his views whether in the voice of a character or through the narrator's voice. Therefore, we hear the Field Assistant Dilip Saikia, remembering the words of his officer:

> Our mission is not merely to build a road through the hills; we should also work towards making our way through the mental barrier of the hill tribes ... they look at each other suspiciously owing to their longstanding rivalry. They do not want to mingle with others; they do not want to live together or work together. During the process of road-construction, we will make the different tribes stay together to rid them of their misgivings about each other. We must bring them together and cement a bond of trust between them (*Silent Lips Murmuring Hearts*: 34)

This is not only the realization of a character of a novel. Instead it is the novelist's inner urge to bring the different tribal people together under the same roof.

It is the ideology of their society that makes the Nyishis believe that they are ethnically superior. Similarly, the Sherdukpens believe themselves to be more "refined" and "polished" than the other tribes. This ideology indoctrinated from his childhood makes the writer say:

> In most cases, after the first ritual that is, after the groom's family members forcefully make off with the bride to take her to the groom's house, the groom's family keeps her forever. Rinchin's family, the Thongdoks, not only belonged to the highest class, but also were one of the most aristocratic families of the village. Consequently, it would have been a matter of risking their family's honour if they were to detain the bride right away after the first abduction (*Silent Lips Murmuring Hearts*: 40).

The conflict in Rinchin's mind regarding his relationship with Yama is a result of his ideas of reputation and family name. 'Ideology' does not allow a person to move from his/her pre-established 'subject-position'.

Rinchin greatly values the image that society grants him. He is a reputed man of his family and, therefore, the society imposes some dos and don'ts upon him. Therefore, to love and marry a girl from an "inferior" tribe might jeopardise the position that he enjoys. One may say that the ideas that the writer has indoctrinated as a social being does not let him show the union of Rinchin and Yama in spite of their intense love. As ideology decides the way in which one perceives the world and its workings.

Similarly, at one point in the novel, the writer himself seems to be expressing the view that Sherdukpen dance forms like Yak, Ajyi Lamu and Tsange are "richer" than the Nyishi dances and Yama and Yado are overwhelmed by the richness of these tribes:

> It seemed to the Nyishi girls that in comparison to the Sherdukpens, their own dances were quite ordinary (*Silent Lips Murmuring Hearts*: 171).

Dominance of patriarchy in the society is an important issue of the novel. Apparently, it seems that tribal societies are generally liberal in terms of issues related to women and marriage. But this does not really appear to be so in the world of *Mouna Ounth Mukhar Hriday*. The women are free to go outside the house to work in the field, and to collect firewood and vegetables. But they do not really seem to play any significant role while taking any important decisions about themselves or about any matters of the house. One can see the influence of patriarchy upon Takar, a character of the novel, who after getting money from his employer happily imagines that he can now 'buy a girl':

> He could buy a female mithun in exchange for a sack of salt ... Once he had a large number of mithuns, he could marry a number of girls. He would be known in the village as Takar the Rich or Takar the Great with ten wives (*Silent Lips Murmuring Hearts*: 91).

This gives an idea of the patriarchal nature of the society for Takar talks of "buying" and "possessing" a number of wives.

Mouna Ounth Mukhar Hriday is the expression of human values, the "simplicity" of the tribal mind, the feeling of love and friendship, and the community feeling in the minds of the tribal people. Yeshe Dorjee Thongchi has introduced the readers not only to the lexical and cultural world of his people but to the inner human feelings in them.

Mouna Ounth Mukhar Hriday is an artistic fiction through which we are introduced to human experiences of a particular age and time. The fiction is not just a reflection of social practices but is an intervention of the social practices, human values and potentiality. These, in a way, shape the thinking of the people and at the same time these are expressed in a work of art, whether literary or otherwise. The community feeling and love and friendship towards others allow Tadak to share a kill (goat) with Rinchin in spite of their difference in their language, dresses, food-habits and rituals. Even though strictly confined to his own community and its marriage-rituals, Tadak can understand Rinchin's love for Yama and expresses his inability to let them be united. This is because of their dominant social norms due to the difference in their languages or customs and ritual. He expresses his helplessness:

> I know that Yama is head over heels in love with you. But you can't deny the fact that we are from different tribes and we follow different rituals ... Yama is another man's wife. Even if you get married, Yama's husband will never spare you. He will slaughter both of you (*Silent Lips Murmuring Hearts*: 124).

Due to all these reasons, Tadak advises Rinchin to forget Yama. The same Tadak who is so obstinate about allowing Yama to marry Rinchin, accepts the latter's request for staying back till the arrival of the governor as Rinchin wants to see Yama for some more days. Tadak's simple, tender heart cannot be cruel to Rinchin. The author has shown the simplicity of Tadak, the friendship between Rinchin and Tadak, and their parting from another with heavy hearts.

In *Mouna Ounth Mukhar Hriday* we see that Rinchin goes hunting on several occasions. Rinchin is a Sherdukpen and the Sherdukpens are Buddhists by religion. It would be important to note here that hunting is not a permissible act for Buddhists. Rinchin not only hunts with traditional weapons but even goes to the extent of procuring a license for a gun. Rinchin's act of hunting is an instance of departure from religious idealism.

Yeshe Dorjee Thongchi deliberately shuffles between the position of an insider and that of an outsider in *Mouna Ounth Mukhar Hriday*. While most of the time he appears to be an insider with a deep sense of affiliation with Arunachalee society, occasionally he appears as a detached observer. We see the writer adopt the latter perspective when he tries to appreciate a haunting melody of a song sung by a girl from the Nyishi camp through the words of Dilip Saikia:

> Even though he could not understand the language, he felt the lilting song have a sort of magical effect. It held the same kind of appeal as the body of an unknown girl of an unknown tribe (*Silent Lips Murmuring Hearts*: 18).

It is the custom of the Monpas to slice dead bodies into a hundred and eight pieces before throwing these pieces into the river. *Saba Kota Manuh* is based on this tradition of the Monpa tribe. The readers are not always conscious of the complex and inexplicable relation between a work of literature and the ideological world the writer and his characters inhabit.

It has been already mentioned that the story of *Saba Kota Manuh* revolves round the typical Monpa custom of slicing dead bodies into a hundred and eight pieces. As such, the action of the story revolves round this custom along with the mention of some other traditions, rituals and beliefs. The customs and rituals of a society is nothing but the belief in some ideological order one internalizes as a social being. These are just social constructs and the member of a particular society is also product of such social constructs. Therefore, Ao Thampa, the man indulging in the job of slicing the dead bodies into hundred and eight pieces, wholeheartedly believes that it is a sacred matter to slice the dead bodies and so, he believes to earn salvation, one should not charge anything for that work. Ideology is so deeply rooted in him that even if he wants to leave this job, he is unable to do so because he firmly believes that it is a sin to say 'no' to such a divine calling:

As there was a scarcity of men who could slice dead bodies, he had no way out but to take it up as a social and religious responsibility which he always carried out dutifully (*Saba Kota Manuh*: 114).

It is this sense of ideology that compels Ao Thampa to consider it a grave sin to think fondly of the great Rimpoche Noraljom Aane Sangge (*Saba Kota Manuh*: 181). Aane Sangge's words, "Cremate the deadbody respectfully. See to it that the dead body is cut into the stipulated number of parts" (*Saba Kota Manuh*: 219) is an indication of her ingrained belief in her tradition. We can draw a parallel between this and the ideology that the author has indoctrinated as a Buddhist by religion. It is the author's Buddhist background which enables him to show Aane Sangge's father to be fortunate enough to die just after his daughter, who is a great Rimpoche, feeds him a morsel of food (*Saba Kota Manuh*: 218). The author's deep faith in his religion makes him announce through the mouth of Rijomba's father:

> It is the result of the good work done by me and your mother that we have earned God's blessings to be the parents of a Rimpoche (*Saba Kota Manuh*: 213).

It is perhaps because of the 'ideology' that the author has indoctrinated as a Buddhist that restricts him from depicting a usual sort of passionate love-affair between Rijomba, the Rimpoche and Ao Thampa, the man who slices the dead bodies. There is no real union of the two in the novel. The love that they share is something spiritual; the lovers attain peace and bliss in spite of their separation at the end.

A sense of spirituality and respect has given way to his physical desire for Rijomba as she is now a great Rimpoche, Noraljom Aanne Sangge. He loves her wholeheartedly but this love is much beyond the bounds of the material world (*Saba Kota Manuh*: 200).

On the other hand, Rijomba's confession to Ao Thampa is a reflection of the conflict in her own mind: "I do not like this life. I can't be free, there are restrictions on everything that I do and think ... What is the use of such devotion till I die if I have to be born again as a Rimpoche?" (*Saba Kota Manuh*: 165).

It is the social norm and ideology of her society that prevents Anne Sangge from leaving her religious life and starting a new familial life with her lover. But the irony is that it is ideology itself which has established the position of the author as an administrative officer rather than as a Buddhist. And the man in him, the author in him allows him to be more of a human being than a Buddhist; therefore, he is free enough to write a story based on the love-affair between a *thampa* and a great Rimpoche. One can infer from this that Thongchi is sufficiently "autonomous" as a writer to depict a love affair between a Rimpoche, who is supposedly free from the worldly pleasures and pain and Ao Thampa, a man who, in spite of being involved in the sacred task of slicing dead bodies, is being hated by others.

If we try to locate the source of Yeshe Dorjee Thongchi's material in *Saba Kota Manuh*, we are provided a clue in the preface of the novel,

> I have felt the urge to write this story since the time I was in school. Once, on the way to my friend Dechin Rokpu's village during my school days, I saw a scene of a dead body being sliced into pieces and drifted in the river ... When serving as a government officer in Tawang I happened to see a scene of dead bodies being sliced and drifted in river once again; on this occasion my daughter Dolam Wangjom was there with me and she asked me to write a novel on this custom (*Saba Kota Manuh*: 5).

There is no doubt that the writer got the material for his work from his personal experience. In an article entitled 'Mor Sahitya Charchar Bisoye Mor Boktobyo' ('My Views About My Literary Works') in the Assamese magazine *Satsori,* Thongchi writes about another lasting image in his mind that was also a source of inspiration in writing this novel:

> In our childhood days, a lame Rimpoche had crossed by our village. She was a middle-aged beautiful Rimpoche who was riding a horse ... the respect that the villagers showed towards her and the struggle between them for getting her blessings remained forever in the innermost recesses of my heart (Thongchi, 2006: 10).

Dalai Lama's visit to Tawang in the 1950s was a matter of great political and historical significance. Another incident of great political and historical import against which the novel is set is the Chinese aggression of 1962. At that point of time, Thongchi was obviously greatly influenced by these two events, so much so that he wrote *Saba Kota Manuh* - a novel about the relationship between a lame Rimpoche and a man who slices dead bodies against the background of the war of 1962. The contrast between the soothing effects of Dalai Lama's visit and the ravages of the Chinese aggression obviously had an effect upon Thongchi.

It is noteworthy that Ao Thampa, the man who slices dead bodies into pieces and considers it to be a sacred work is shown to be dirty and foul smelling, - a man whom the villagers keep away from except at the time of someone's death. In spite of the character's deep faith in the system of slicing dead bodies as a sacred work, the author is unable to portray Ao Thampa as a "normal", "natural" human being with the same human feeling and emotions. Instead, Ao Thampa appears to be an unstable man who behaves whimsically: "He is unable to do anything except slicing the dead bodies. He is unwilling to do anything else. He is lazy and becomes a drunkard" (*Saba Kota Manuh*: 35).

The irony is that the custom of slicing the dead bodies is considered to be a holy work, but the man involved in it is not a "normal" human being, not happy in his familial life and moreover his own daughter is physically disabled. The author makes Ao Thampa remember his mother's words, "His mother warned him not to be a *thampa*, a man who slices dead bodies. The belief that through the sacred work of slicing dead bodies one can earn salvation, is proven to be untrue in his life" (*Saba Kota Manuh*: 41). This is perhaps an expression of the author's sense of skepticism in such traditional belief.

In 'My Views about My Literary Works' the author talks about the deep sense of pain that he felt during the writing of *Saba Kota Manuh* (Thongchi, 2006: 10). Thongchi does not elaborate the reasons behind the pain that he experienced while writing the novel. But we have reasons to believe that this pain could be the result of Thongchi's sense of scepticism regarding a custom prevalent in a tribe very close to his own.

The sense of pain that he experienced could also have something to do with the kind of objectivity with which he is able to depict the customs of a society so close to his own. The vivid and realistic description of the custom of slicing dead bodies into pieces evokes a sense of horror and repulsion that can only be explained by the detachment with which the author is able to look at his own society. Towards the end of *Saba Kota Manuh*, the *thampa* gives up the work of slicing dead bodies; he becomes "clean", "rational" and "natural", and in the author's own words a "good" human being (*Saba Kota Manuh*: 126).

Conclusion

Lummer Dai and Yeshe Dorjee Thongchi have carved a niche for themselves in the history of the Assamese novel by their distinctness of approach, their deep empathy for the tribal way of life, and above all, their critical eye. Both the writers' works are marked by certain distinct characteristics of the tribes to which they belong. But while being deeply rooted in their own milieu, they are also conscious of the social process of the formation of tribal identity as a whole. This consciousness of the authors manifests itself in their concern for the growth and development of a pan-Arunachalee identity. The writings of these authors lend itself quite easily to a New Historicist analysis as they are so expressive of the complex multicultural set-up of Arunachal Pradesh.

Today, following the advent of New Historicism as a critical theory, the relationship between the author and his/her literary work is far more complex than what was thought to be the case during earlier forms of historicism. The views and practices of the New Historicists differ markedly from those of former scholars who had used political and intellectual history as a "background" to account for the characteristic subject-matter of literature at a particular time and place. We can no longer simply talk in terms of a writer's personal life and beliefs asserting itself on his/her work directly. Rather, we have come to understand and appreciate the complex network of power relations and ideology at work in the process of any literary production.

The very understanding of the concept of the writer's subjectivity has altered considerably - subjectivity is no longer a simple matter of the writer pouring his/her heart out in literature. New Historicism as a critical theory owes a lot to poststructuralists and structuralist Marxists like Louis Althusser who are suspicious of the very idea of 'self'. So, instead of the idea of 'self', we look at the author, reader and characters as 'subjects' produced by the dominant ideology. New Historicists do not try to show merely that the literary text reflects a historical event but make us see any social event as a literary text and the literary text as a social event.

In the epilogue of *Renaissance Self-Fashioning* (1980) Stephen Greenblatt says that, in the course of writing the book, he lost his initial confidence in "the role of human autonomy", for "the human subject itself began to seem remarkably unfree, the ideological product of the relations of power in a particular society" (quoted in ibid. 1993: 250). A major area of contest among New Historicists is the extent to which an author, despite being a "subject" produced and positioned by the play of power and ideology within the discourse of a particular era, may be able to retain some scope for individual initiative and "agency" (ibid. 1993: 250-51).

New Historicists emphasize the importance of ideology, the socio-political disposition, unknown to an author himself that governs his/her work. The typical way in which a New Historicist would go about his/her task of reading Shakespeare, for example, would be to initially gather the sundry facts available on his life and try to connect these to his plays. While the older historicist would have been contented with this, the New Historicist would go on to use the plays as history and take help of these plays to throw more light on Shakespeare's life.

Lummer Dai wrote his first three novels before he joined government service. It is interesting to note the subtle differences between these and the other two novels that he wrote after joining government service. The unmistakable streak of reformist zeal that is clearly discernible in the later works of Lummer Dai has clearly a lot to do with the social role that he was playing in those days. *Kanyar Mulya* was published in the days when he was serving as a Public Relations Officer and it was almost "obligatory" on his part to work towards a reformation of Arunachalee society. Gumba constantly speaks out against the practice of bride-price in the novel *Kanyar Mulya* even before she is aware of the fact that she herself is a victim of the same practice. Quite clearly, the motive for Gumba's protest comes from without, - it comes from none other than Dai himself. However, this is not to say that Dai's reformism is in any way a conscious effort; in fact, the reformism is there in spite of Dai.

Upar Mahal was written when Dai was serving as the Director, Information and Public Relations and was published in 2002, the year he breathed his last. Holding a very important position of Director, IPR, Dai had the experience of observing the scandalous activities of contractors, politicians and bureaucrats in the capital town of Itanagar from close quarters. This led him to say in an interview that he was "angry" and "apprehensive" (*Dainik Asom*, 1999, 22nd March) on seeing the depravity in the field of politics. Therefore, we have an anti-hero in the form of Jalam in *Upar Mahal*. Jalam is a depraved contractor who finally rises to the post of Home Minister and the sharp rise in his material prosperity is only paralleled by a simultaneous decline in moral values.

In a posthumous biodata of Lummer Dai circulated by Nani Dai, his wife, the late writer is described as being "... a great champion of social reformation. Late Dai, through his thought provoking writings, had always advocated elimination of social evil practices like polygamy, child marriage and forced marriage, slavery etc. and social emancipation of tribal women" (N. Dai, 2003).

The urge to reform is of course also discernible in Dai's earlier novels. But one cannot deny the fact that there is a remarkable difference in the reformist zeal of *Kanyar Mulya* and the feeling of contentedness in just being able to portray life as it is in *Paharar Xile Xile* and *Prithivir Hanhi*. In *Paharar Xile Xile*, Jalam's father is entirely unhappy about the marriage between Bati and his son because she happens to be a slave's daughter. This situation in the novel provided an opportunity to Dai to show Bati as a victim of the social evil of slavery as he made Gumba protest against child marriage in *Kanyar Mulya*. There are many misfortunes that befall Jalam's family towards the later part of the novel but none of them can be necessarily explained as the consequence of castigating Bati as the daughter of a slave. It is shown to be the consequence of Jalam's disbelief and punishment of Bati, not on the grounds of her being a slave's daughter. Gumba resists the social structure and survives while Bati perishes in the process.

Prithivir Hanhi was Dai's first novel to introduce us to the transition taking place in Arunachalee society, the second being *Upar Mahal*. Kardug disregards the *kebang* only for his selfish motive and beats Liyi to death right in front of the members of the *kebang*. In Misra's words,

> From the very first page where the author presents an early morning scene inside a *musup* (bachelor's dormitory), to the scenes inside a Chang where the newly wed share the sleeping space with all the other members of the bride's family, to the last scene where Liyi boldly speaks out against Kardug's immoral behaviour in a village *kebang* – the novel leaves the indelible impression that the author has portrayed the cultural life of his own community in an inimitably realistic manner (2001: 34).

The character of Kardug, the protagonist, problematizes the myth of the naivety of tribal people. Dai portrays Kardug as an extremely complex character whose cruelty and material ambition leads him to unscrupulously dupe his friend Lib and desert him in Dibrugarh.

Mon aru Mon expresses a kind of conflict, not only of the characters' minds but of the mind of the writer too. The narrator's direct comment, comments through the voice of the characters of the novel, manner of describing the folk beliefs and even the footnotes definitely reveal many things not only about the societies but of the writer and his attachment to the society he is writing about.

If Lummer Dai's novels hold up before us a living picture of Adi society, Thongchi's novels transport us to the world of the Sherdukpen and the Monpa societies. Thongchi, unlike Dai, is married to a member of a different tribe from his own. Mrs. Sonamati Welly, Thongchi's wife, is a Nyishi. The union of the Sherdukpen lover Rinchin Norbu and his Nyishi beloved Yama Natung remains a dream in *Mouna Ounth Mukhar Hriday*. Intertribal marriage is not acceptable, is in fact an impossibility in the world of Rinchin, Tadak and Yama. But in the real life of the author this dream of intertribal marriage becomes a reality. One may say that *Mouna Ounth Mukhar Hriday* is an autobiographical novel having similarities with Thongchi's personal life but with an altered conclusion. The fact that Rinchin and Yama fail to be united at the end of the novel could be an expression of the feeling of guilt of the author at the subconscious level regarding his own marriage with a member of another tribe.

Mouna Ounth Mukhar Hriday is different from Thongchi's other novels in the sense that it is not confined to the customs and rituals of one single tribe. Thongchi skillfully traces the initial feeling of suspicion of the tribes towards one another and tries to bring about a feeling of unity and camaraderie between them. The novel was written in 1990-91 when Thongchi was serving as a senior administrator of the state and in 1992 he was nominated as an IAS. As an administrator it was his duty to foster a spirit of unity amongst the tribes of Arunachal Pradesh. Dilip Saikia, the field assistant of *Mouna Ounth Mukhar*

Hriday, is a mouthpiece for Thongchi to express his opinions regarding the need for the tribes to get rid of the misgivings about each other and be united.

Yeshe Dorjee Thongchi was born in the Jee *basti* of West Kameng District in the year 1952 and later rose to the position of an IAS officer. As he moved ahead in the journey of life, he probably lost touch with many near and dear ones. But, he believes in getting along in life with great optimism. Thongchi seems to believe in making new beginnings in life and sees hope and light in the onward journey of life - breaking free of the stifling customs and rituals to move ahead in life without being disrespectful of his origin. This explains the repeated use of the journey metaphor in Thongchi's novels. In addition to *Mouna Ounth Mukhar Hriday* which ends with the onward journey of Rinchin and Tsering Jangmu in one direction and Yama in another, *Sonam* also ends with the journey of Lobjang and the others towards Tawang:

> This was a journey to keep the world alive, a journey to build up a happy family, a journey to build the future forgetting the past (*Sonam*: 127).

Lingjik ends with the breaking of the traditional Sherdukpen *lingjik*, again symbolizing a break with the past and the journey toward the future with new hope and vigour. The character of Prem Tasi in *Lingjik* is, to a great extent, autobiographical. Prem Tasi is an Extra Assistant Commissioner who comes from the same kind of familial set-up as Thongchi. They are both first generation literates and both rise to the level of administrative officers.

Lingjik tells us about the reconciliation of Thongchi/Prem Tasi with his own root. Prem Tasi, after his new-found status as an administrative officer, looks down upon the traditional belief system of his society. His education makes him think lowly of his own society until he comes to the realization that

> He did not know that every society had its own

> characteristics, values and strong foundation. It is these characteristics and values that impart hope to the society and this hope of society is beautiful. So, no society in the world was better or inferior than any other (*Lingjik*: 132).

Both *Lingjik* and *Sonam,* his first novel, are firmly rooted in certain customs of Sherdukpen and Monpa societies respectively. Being born to a Sherdukpen society which is very close to the Monpa society, Thongchi has observed the Brokpas from close quarters. In the first edition of *Sonam,* he was happy to portray the Brokpa society realistically. But in the new edition of the novel, he felt the need for extending it - Lobjang repents his past deeds and decides to get married to Sonam's sister Tsering Doima. At the end of the novel, Lobjang and Tsering Doima decide to go to Tawang in the hope that they will no longer live as Brokpas and follow the practice of *khor depka.* One thus discovers a new urge in Thongchi to reform society in *Sonam.*

Saba Kota Manuh is Thongchi's latest novel published in the year 2004. Thongchi had seen a lame reincarnated Rimpoche when he was a child and he retained the image in his mind. The character of Aane Sange is quite clearly modeled on the Rimpoche that Thongchi had seen as a child. It is quite possible that Thongchi unconsciously felt attracted to the Rimpoche as a child who was at that time blissfully ignorant of the restriction on such a thought imposed by the adult Buddhist society around him. As an adult such a thought is also clearly not acceptable to Thongchi and therefore he develops an almost spiritual love affair between Aane Sange and Ao Thampa in the novel, effecting a displacement of his own feelings for the Rimpoche to Ao Thampa.

The clinical exactness with which Thongchi describes the activities of Ao Thampa (slicing the dead bodies) clearly shows that he is here looking at the Monpa society quite objectively. He views this society with the kind of detachment that allows him to describe the scenes of horror and repugnance very minutely. Thongchi has journeyed a long way from his society of origin, a distance sufficient for him to feel repelled by the act of slicing dead bodies. Thongchi was surprised to find his own daughter being unmoved by the same scene:

The fact that my adolescent daughter was able to look at the scene without any horror or repulsion surprised me (Thongchi, 2006: 10).

In *Saba Kata Manuh* there are descriptions of the death of Monpas in the earthquake of 1950, the transformation of administrative power of Tawang from Tibetan government to Indian government in 1952, the Chinese aggression and Dalai Lama's entering India through the district of Tawang, the *kalachakra* puja in Dirang by the Dalai Lama, the death of Monpas carrying luggage to the frontier during the war etc.

While analyzing Lummer Dai and Yeshe Dorjee Thongchi's novels drawing on New Historicist perspectives, many undiscovered facts regarding their lives and writings emerge. Through the use of New Historicism I have analyzed how Lummer Dai and Yeshe Dorjee Thongchi's writings can be most fruitfully analyzed to look at their deep-rooted connection with the society of Arunachal Pradesh. And the myth of placing the literary text in isolation, setting it free of the writer and his/her background is shattered, thereby establishing a relation between the writer and the man who writes.

To conclude, Carey Snyder's reading of Chinua Achebe's *Things Fall Apart* may be fruitfully used as an analogy to understand the divided subjectivities of Thongchi and Dai. Like Achebe, Thongchi and Dai live in "divided worlds" and this in fact, becomes the enabling condition of their art (2010: 181). Like skilled ethnographers, Thongchi and Dai straddle between their native and acquired traditions. They are not out to "salvage" a "vanishing" culture out of "disinterested intellectual curiosity", but rather, like Achebe, they are "cultural nationalist[s] interested in recuperating" their cultures (Snyder, 2010: 185).

Bibliography

Select Novels:

Dai, Lummer. *Paharar Xile Xile*. Guwahati: Asom Prakashan Parishad, 1961.

——. *Prithivir Hanhi*. Guwahati: Asom Prakashan Parishad, 1963.

——. *Mon aru Mon*. Guwahati: Dutta Baruah and Co., 1967.

——. *Upar Mahal*. Guwahati: Bani Mandir, 2003.

——. *Kanyar Mulya*. Guwahati: Bani Mandir, 1978. (First Edition). 2004 (Second Edition).

Thongchi, Yeshe Dorjee. *Sonam*. Guwahati: Parbati Prakashan, 1981. (Second Edition). Nalbari: Journal Emporium. Translated into English by Mridula Borooah. Guwahati: Spectrum Publications. 2005.

——. *Lingjhik*. Guwahati: Dutta Baruah and Co., 1983.

——. *Mouna Ounth Mukhar Hriday*. Guwahati: Banalata, 2001. Translated into English by Debarshi Prasad Nath. New Delhi: Sahitya Akademi. 2010.

——. *Saba Kota Manuh*. Guwahati: Banalata, 2004.

Literary Sources (English):

Abrams, M.H. *A Glossary of Literary Terms*. Bangalore: Prism Books. 1993.

Ahmed, Ishteaque Syed. 'Nocte' in *People of India: Arunachal Pradesh.* (ed.) Parul Dutta & Syed Ishteaque Ahmed. Vol. XIV. Calcutta: Seagull Books. 1995. pp 281 - 287.

——. 'Aka' in *People of India: Arunachal Pradesh.* Vol. XIV. (ed.) Parul Dutta & Syed Ishteaque Ahmed Calcutta: Seagull Books. 1995. pp 135 - 140.

Allen, Walter. *The English Novel.* London: Pelican Books. 1954.

Althusser, Louis. 'Ideology and the State'. in *Modern Literary Theory: A Reader.* (ed.) Philip Rice and Patricia Waugh (trans.) B. Brewster. London: Edward Arnold. 1992. pp. 54-62.

Bagchi, R. N. 'Socio-Cultural Profile of the Monpas of Kalaktang Areas' in *Aspects of Culture And Customs of Arunachal Pradesh* (ed.) P.C. Dutta & D. K. Duarah. 1990. pp 37 - 51.

Banerjee, Bikash. 'Tribal Village Council of A.P: Khamba' in *Tribal Village Councils of Arunachal Pradesh.* (ed.). B.B Pandey, D.K. Duarah, & N. Sarkar.Itanagar: Directorate of Research, Government of Arunachal Pradesh. 1999. pp285-293.

Barry, Peter. *Beginning Theory: An Introduction to Literary and Cultural Theory*. Manchester: Manchester University Press. 2002.

Barua, B.K. *Assamese Literature.* Bombay: International Book House. 1941.

——. *Modern Assamese Literature.* Gauhati: Lawyers Book Stall. 1957.

Barua, S.K. 'Role of Dormitories in the Social and Political Life of the Wanchos' in *Aspects of Culture And Customs of Arunachal Pradesh.* (ed.). P.C. Dutta & D. K. Duarah. Itanagar: Directorate of Research, Government of Arunachal Pradesh. 1990. pp. 75 - 82.

Barua, S. 'Zakhring' in *People of India: Arunachal Pradesh.* Vol. XIV. (ed.) Parul Dutta & Syed Ishteaque Ahmed. Calcutta: Seagull Books. 1995. pp. 331-343.

Basu, A., Dasgupta, B.K. & Jayanta Sarkar. (ed.). *Anthropology for North East India: A Reader.* Kolkata: Indian National Confederation and Academy of Anthropologists, Indian Anthropological Society and National Museum of Mankind. 2004.

Behera, M.C. & N.C. Roy. (ed.). *Trends in Agrarian Structure in the Hills of Northeast India.* New Delhi: Commonwealth Publishers. 1997.

Beteille, Andre. *Society and Politics in India: Essay in a Comparative Perspective.* Oxford: OUP. 1999.

Bhagawati, A.C. 'Tribal Transformation in Assam and North-East India: An Appraisal of Emerging Ideological Dimentions' in *Tribal Transformation in India Vol. 3: Ethnopolitics and Identity.* (ed.) B Choudhury Inter India Publication. 1992. pp 487 - 509.

'The Tribe as a Social Formation: The Case of the Tangsa of Arunachal Pradesh' in *Anthropology for North East India: A Reader.* (ed.). A. Basu, B.K. Dasgupta, & Jayanta Sarkar. Kolkata: Indian National Confederation and Academy of Anthropologists, Indian Anthropological Society and National Museum of Mankind. 2004. pp 177 - 187.

Bhattacharjee, S. *Politics of Frontier State: Arunachal Pradesh.* New Delhi: Omsons Publication, 1994.

Bhattacharya, Kishore. "The Interpretation of Folktales of North-East India and the Value of Content Analysis" in *Folk Culture and Oral Literature from North-East India.* (ed.) Tamo Mibang & S. Chaudhuri 2004. pp 25 - 31.

Bhattacharya, R.K. & Jayanta Sarkar. (ed.). *Anthropology of B.S. Guha.* Calcutta: Anthropological Survey of India. 1996.

Blackburn, Stuart. 'Tribal Transitions: Cultural Innovation in Arunachal Pradesh, India. Challenging the Romance of Tribal Culture' in *Indian Folklife, Vol- 2, Issue 1,* Chennai: National Folklore Support Centre, July 2002. pp. 14 - 17.

Blamires, Harry. *A History of Literary Criticism.* New Delhi: Macmillan India LTD, 2000.

Bodong, Yirang. Gindu Borang, & Girin Tamuli. (ed.) *Silifor Review, Vol. II, No. II.* (ed.) Pasighat: Siang Literary Forum. 2006.

Borang, Kaling. 'My Pyaayi - The Feather on the Hat' in *Silifor Review, Vol. II, No. II.* 2006. pp. 2 - 8.

Borgohain, H. 'Customary Land Law of the Adis' in *Tribes of Arunachal Pradesh: History and Culture.* (ed.) Bibhash Dhar. & P.C. Coomar. Delhi: Abhijeet Publications. 2004. pp. 93-101.

Borpujari, H. K. *Problems of the Hill Tribes of North East Frontier.* Gauhati: LBS, 1970.

Borua, H. *Assamese Literature.* New Delhi: NBT. 1965.

Bove, Paul A. 'Discourse' in *Critical Terms for Literary Study.* (ed.). Frank Lentricchia & Thomas McLaughlia. Chicago: University of Chicago Press. 1995. pp 50 - 65.

Chaudhuri, Sarit. 'Folk Literature in the Context of a Pluricultural State: Some Observation on the Emerging Realities of Arunachal Pradesh' (Unpublished). Paper presented in the Sahitya Akademi sponsored seminar 'Folk Literature - Voice of the People: Continuity and Change' in Tezpur University on the 28th February and 1st March 2007.

Chaudhury, B. (ed.). *Tribal Transformation in India Vol. 3: Ethnopolitics and Identity.* New Delhi: Inter India Publication. 1992.

Choudhury, S. Dutta. *Arunachal Pradesh District Gazetteers: East Siang and West Siang Districts.* Itanagar: Government of Arunachal Pradesh, 1994.

Chowdhury, J.N. *Arunachal Panorama.* Itanagar: Directorate of Research, Government of Arunachal Pradesh. 1979.

——. *Arunachal Pradesh: From Frontier Tracts to Union Territory.* New Delhi: Cosmo Publication. 1983.

Chowdhury, S. 'Tribal Village Council of A.P: Khampti' in *Tribal Village Councils of Arunachal Pradesh.* (ed.). B.B Pandey, D.K. Duarah, & N. Sarkar. Itanagar: Directorate of Research, Government of Arunachal Pradesh. 1999. pp 315-329.

Clifford, James. *On the Edges of Anthropology.* Chicago: Prickly Paradigm Press. 2003.

Das, A. R. 'Tagin' in *People of India: Arunachal Pradesh.* Vol. XIV. (ed.). Parul Dutta & Syed Ishteaque Ahmed. Calcutta: Seagull Books. 1995. pp. 308 - 312.

Das, S.T. *Tribal Life of North-eastern India.* Delhi: Gyan Publishing House. 1985.

Das, Veena. (ed.) *Oxford India Companion to Sociology and Social Anthropology.* New Delhi: OUP. 2003.

Dawar, J.L. *Cultural Identity of Tribes of North-East India.* New Delhi: Commonwealth Publishers. 2003.

——. 'Adi Intellectuals' Perception of Cultural Identity' in *Studies in the History, Economy and Culture.* (ed.) S. Dutta. Itanagar: Himalayan Publishers. 2000. pp 408 - 421.

Deka, Umesh. *Post-War Assamese Novel.* Guwahati: Chandraprakash. 1991.

Deuri, R.K. *The Solungs.* Shillong: Directorate of Research, Government of Arunachal Pradesh. 1982.

——. *Festivals of Kameng.* Shillong: Directorate of Research, Government of Arunachal Pradesh. 1983.

Dhar, Bibhash. 'Bangni' in *People of India: Arunachal Pradesh.* Vol. XIV. (ed.). Parul Dutta & Syed Ishteaque Ahmed Calcutta: Seagull Books. 1995. pp. 147 - 150.

Dhar Bibhash. 'Khowa' in *People of India: Arunachal Pradesh.* Vol. XIV. (ed.) Parul Dutta & Syed Ishteaque Ahmed. Calcutta: Seagull Books. 1995. pp. 185 - 189.

Dhar, Bibhash. & P.C. Coomar. (ed.) *Tribes of Arunachal Pradesh: History and Culture.* Delhi: Abhijeet Publications. 2004.

Dutta, D.K. 'An Ethnographic Note on the Lisus' in P.C. Dutta & D. K. Duarah. (ed.). *Aspects of Culture And Customs of Arunachal Pradesh.* Itanagar: Directorate of Research, Government of Arunachal Pradesh, pp. 14 - 30. 1990.

Dutta, Parul & Syed Ishteaque Ahmed (ed.). *People of India: Arunachal Pradesh.* Vol. XIV. Calcutta: Seagull Books. 1995.

Dutta, Parul. 'A Note on the People of Arunachal, Their Culture and Matters of Anthropological Interest' in *Tribes of Arunachal Pradesh: History and Culture.* (ed.) Bibhash Dhar & P.C. Coomar. Delhi: Abhijeet Publications. 2004. pp 5 - 14.

Dutta, Parul. 'The Wanchos of Arunachal Padesh: Their Cultural and Historical Background' in *Tribes of Arunachal Pradesh: History and Culture.* (ed.) Bibhash Dhar & P.C. Coomar. Delhi: Abhijeet Publications. 2004. pp 27 - 47.

Dutta, P.C. & D. K. Duarah. (ed.). *Aspects of Culture and Customs of Arunachal Pradesh.* Itanagar: Directorate of Research, Government of Arunachal Pradesh. 1990.

Eagleton, Terry. *Literary Theory: An Introduction.* New Delhi: Maya Publishers. 2000.

—— . *Marxism and Literary Criticism.* London: Routledge Classics. 2002.

Elwin, Verrier. *Myths of the North-East Frontier of India.* Itanagar: Directorate of Research, Government of Arunachal Pradesh. 1958.

——. *A Philosophy for NEFA.* Itanagar: Directorate of Research, Government of Arunachal Pradesh. 1959.

—— . *A New Book of Tribal Ficiton.* Itanagar: Directorate of Research, Government of Arunachal Pradesh. 1970.

Erikson, Paul A. & Liam D. Murphy. *A History of Anthropological Theory.* Toronto: Broadview Press. 1998.

Felluga, Dino. 'General Introduction to New Historicism' in *Introductory Guide to Critical Theory*. **Nov. 28, 2003**. Purdue U. April 8th 2007.

<http://www.purdue.edu/guidetotheory/newhistoricism/modules/introduction.html>.

Forster, E.M. *Aspects of the Novel*. London: Edward Arnold. 1949.

Foucault, M. *Power/Knowledge*. (ed.). Colin Gordon. (trans.) Colin Gordon, Leo Marshall, John Mepham, Kate Soper. Hervester: Hervester Press. 1980.

——. *Politics, Philosophy and Culture*. (Trans.) Alan Sheridan et al. London: Routledge. 1990.

Frow, John. *Marxism and Literary History*. Oxford: Basil Blackwell. 1988.

Gallagher, Catherine. 'Marxism and the New Historicism' in *The New Historicism*. (ed.) Aram H. Veeser. London: Routledge. 1989. pp 37 - 48.

Geertz, Clifford. *The Interpretation of Culture: Selected Essays*. New York: Basic Books.1973.

Gewertz, Ken. 'Greenblatt teases out a knowable Shakespeare'. Harvard University Gazette, September 30, 2004. 4th November 2006.

http://www.hno.harvard.edu/gazette/2004/09.30/01-shakespeare.html

Gopalakrishnan, R. *The North-East India: Land, Economy and People*. New Delhi: Har-Anand Publication. 1991.

——. *Arunachal Pradesh: Land and People*. New Delhi: Omsons Publications. 1994.

Gohain, Hiren. *The New Millennium Pronouncing Anglo-Assamese Dictionary*. Guwahati: Banalata. 2002.

Goswami, M.C. & P.B. Das. *The People of Arunachal Pradesh: A Physical Survey*. Itanagar: Directorate of Research, Government of Arunachal Pradesh. 1990.

Greenblatt, Stephen. *Renaissance Self Fashioning*. Chicago: Chicago University Press. 1980.

—— 'Towards a Poetics of Culture' in *The New Historicism*. (ed.) Aram H Veeser. London: Routledge. 1989. pp. 1 - 14.

——. *Learning to Curse*. London: Routledge, 1990.

—— *Marvelous Possessions: The Wonder of the New World*. Oxford: Clarendon Press, 1991.

—— . 'Culture' in *Critical Terms for Literary Study*. (ed.) Frank Lentricchia and Thomas McLaughlia. Chicago: University of Chicago Press. 1995. pp 225 - 232.

——. *Hamlet in Purgatory*. New Jersey: Princeton UP. 2001.

——. 'The Death of Hamnet and the Making of Hamlet' in *The New York Review Of Books*. VOLUME 51, NUMBER 16 · OCTOBER 21, 2004.

Grewal, D.S. *Tribes of Arunachal Pradesh: Identity, Culture and Languages*. Delhi: South Asia Publications. 1997.

Harvey, Paul. *The Oxford Companion to English Literature*. Oxford: Clarendon Press. 1946.

Haimendorf, Christoph Von Furer. *A Himalayan Tribe: From Cattle to Cash*. New Delhi: Vikas Publishing House. 1980.

—— . *Highlanders of Arunachal Pradesh: Anthropological Reseacrh in North East India*. New Delhi: Vikas Publishing House. 1982.

James, Henry. *The Art of Fiction*. New York: Vilsin Press. 1951.

Kakati, B.K. *Assamese: Its Formation and Development*. Gauhati: LBS. 1972.

Karotempreal, S. (ed.) *The Tribes of North East India*. Shillong: Venderame Missiological Institute. 1984.

Kavanagh, James H. 'Ideology' in *Critical Terms for Literary Study*. (ed.). Lentricchia, Frank & Thomas McLaughlia. Chicago: University of Chicago Press. 1995. pp. 306 - 320.

Kettle, Arnold. *An Introduction to the English Novel*. New Delhi: UBS. 1996.

Koley, R. N. 'Sherdukpen' in *Tribal Village Councils of Arunachal Pradesh*. (ed.) B.B Pandey, D.K. Duarah, & N. Sarkar. 1999. pp 13 - 39.

Leitch, Vincent B. *Cultural Criticism, Literary Theory, Poststructuralism*. New York: Colombia University Press. 1992.

Lentrichia, Frank. 'Foucault's Legacy: A New Historicism?' in *The New Historicism*. (ed.) Aram H, Veeser. Chicago: University of Chicago Press. 1989. pp 231 - 242.

Lentricchia, Frank & Thomas McLaughlia. (ed.). *Critical Terms for Literary Study*. Chicago: University of Chicago Press. 1995.

Lukes, Steven. *Marxism and Morality*. Oxford: OUP. 1988.

Majumdar, D.N. & T.N. Madan. *An Introduction to Social Anthropology*. New Delhi: National Publishing House. 1988.

Malinowski, B. *Magic, Science and Religion and Other Essays*. New York: Doubleday Anchor Books. 1955.

Maitra, Asim. 'Memba' in *People of India: Arunachal Pradesh*. Vol. XIV. (ed.) Parul Dutta & Syed Ishteaque Ahmed. Calcutta: Seagull Books. 1995. pp. 190-195.

Megu, A. 'Tribal Village Council of A.P: Adi (General) in *Tribal Village Councils of Arunachal Pradesh*. (ed.). B.B Pandey, D.K. Duarah, & N. Sarkar. Itanagar: Directorate of Research, Government of Arunachal Pradesh. 1999. pp 211-224.

Mibang, T. 'Adi Society in Pre-Independence Period' in *Studies in the History, Economy and Culture*. (ed.) S. Dutta. Itanagar: Himalayan Publishers. 2000. pp 422 - 427.

Mibang, T. & M.C. Behera. (ed.). *Tribal Villages in Arunachal Pradesh: Changing Human Interface*. Delhi: Abhijeet Publications. 2004.

Mibang, T. & S. Chaudhury (ed.). *Folk Culture and Oral Literature from North-East India.* New Delhi: Mittal Publications. 2004.

Misra, Tilottama. 'The Multi-Cultural Dimensions of the Asamiya Novel' in *Indian Horizons Vol. 48, No. 3.* (Guest ed.) Mrinal Miri. New Delhi: ICCR. 2001.

Misra, U. *India's North East: Quest for Identity.* New Delhi: Omsons Publications, 1988.

Mitchell, W.J.T. 'Representation' in *Critical Terms for Literary Study.* (ed.). Frank Lentricchia & Thomas McLaughlia. Chicago: University of Chicago Press. 1995. pp. 11 - 22.

Mitkong, N. 'Nah' in Tribal Village Council of A.P: Nah' in *Tribal Village Councils of Arunachal Pradesh.* (ed.). B.B Pandey, D.K. Duarah, & N. Sarkar. Itanagar: Directorate of Research, Government of Arunachal Pradesh. 1999. pp 224-237.

Morgan, Lewis Henry. *Ancient Society.* Calcutta: KP Bagghi & Co.. 1877. Revised Edition 1996.

Montrose, Louis A. 'Professing the Renaissance: The Poetics and Politics of Culture' in *The New Historicism.* (ed.) Aram H. Veeser. London: Routledge. 1989. pp 15 - 36.

Moore, Jerry D. *Visions of Culture: An Introduction to Anthropological Theories and Theorists.* Toronto: Alta Mira Press. 2004.

Murfin, R. & Supriya M. Ray. *The Bedford Glossary of Critical and Literary Terms.* Boston: Bedford Books. 1998.

Myers, D.G. 'The New Historicism in Literary Study' in *Academic Questions Vol. 2* Winter: 1988-89. pp. 27-36.

Narain, Iqbal (ed.) *Literature, Social Consciousness and Polity.* New Delhi: Manohar Publications. 1987.

Narayan, Kirin. 'How Native Is a "Native" Anthropologist?' American Anthropologist, New Series, Vol. 95, No. 3 (Sep.1993). pp. 671-686.

Newton, Judith L. 'History as Usual? Feminism and the "New Historicism"' in *The New Historicism*. (ed.) Aram H. Veeser. London: Routledge. 1989. pp 152 - 167.

Niall, Lucy. *Postmodern Literary Theory: An Introduction*. Oxford: Blackwell. 2002.

Norbu, T. 'Tribal Village Council of A.P: Monpa' in *Tribal Village Councils of Arunachal Pradesh*. (ed.). B.B Pandey, D.K. Duarah, & N. Sarkar. Itanagar: Directorate of Research, Government of Arunachal Pradesh. 1999. pp 1-12.

Nyori, Tai. *History and Culture of the Adis*. New Delhi: Omsons Publications. 1993.

Ousby, Ian. *Cambridge Paperback Guide to Literature in English*. Cambridge: CUP. 1996.

Pandey, B.B. *Arunachal Pradesh: Village State to Statehood*. Itanagar: Himalayan Publishers. 1997.

Pandey, B.B., D.K. Duarah, & N. Sarkar. (ed.). *Tribal Village Councils of Arunachal Pradesh*. Itanagar: Directorate of Research, Government of Arunachal Pradesh. 1999.

Panja, Sharmistha (ed.). *Critical Theory: Textual Application*. Delhi: Worldview. 2002.

Pathy, Jaganath. 'Tribal and Indigenous Peoples of the World' in *Tribal Transformation in India Vol. 3: Ethnopolitics and Identity*. (ed.) B Choudhury. New Delhi: Inter India Publication. 1992. pp 13 - 27.

Patterson, Lee. 'Literary History' in *Critical Terms for Literary Study*. (ed.). Frank Lentricchia & Thomas McLaughlia. Chicago: Chicago University Press. 1995. pp 250 - 262.

Robinson, Mairi (ed.). *The Chambers 21st Century Dictionary*. New Delhi: Allied Chambers (India) Limited. 1997.

Roy Burman, B. K. 'Transformation of Tribes and Analogous Social Formations' in. *Tribal Transformation in India Vol. 3: Ethnopolitics and Identity*. (ed.)

B.Chaudhury. New Delhi: Inter India Publication. 1992. pp 28 - 34.

Ryan, Michael. *Literary Theory: A Practical Introduction.* Oxford: Blackwell Publications. 2002.

Sabhapandit, P. *Sociological Study of the Post-war Assamese Novel.* New Delhi: Omsons Publications. 1988.

Saikia, N. *Background of Modern Assamese Literature.* New Delhi: Omsons Publications. 1988.

Sangma, Milton S. *History of American Baptist Mission in North East.* Delhi: Mittal Publication. 1987.

Saraswati, B (ed.) *Tribal Thought and Culture.* New Delhi: Concept Publishing Company. 1991.

Sarkar, N. *Buddhism among the Monpas and Sherdukpens.* Shillong: Directorate of Research, Arunachal Pradesh. 1980.

Sarkar, Subhash C. *Indian Literature and Culture.* New Delhi: D.K. Publishers. 1991.

Sharma, R.R.P. *The Sherdukpens.* Shillong: Directorate of Research, Arunachal Pradesh. 1960.

Sharma, Suresh. *Tribal Identity and the Modern World.* New Delhi: Sage Publications. 1994.

Singh, A.K. & M.K. Jabbi. (ed.). *Tribals in India: Development, Deprivation and Discontent.* New Delhi: Har-Anand Publications. 1995.

Srinivas, M.N. 'Social Anthropology and Literary Sensibility' in *Economic and Political Weekly* Vol. XXXIII, No. 39. (ed.) Krishna Raj. September 1998. pp. 2525 - 2528.

Srutikar, P. K. 'Nishing' in *Tribal Village Councils of Arunachal Pradesh.* (ed.) B.B Pandey, D.K. Duarah, & N. Sarkar. Itanagar: Directorate of Research, Government of Arunachal Pradesh.1999. pp 77- 98.

Subba, T.B. & G.C. Ghosh. (ed.) *The Anthropology of North East India.* New Delhi: Orient Longman. 2003.

Snyder, Carey. 'The Possibilities and Pitfalls of Ethnographic Readings: Narrative Complexity in *Things Fall Apart*' in Harold Bloom (ed.) *Chinua Achebe's* Things Fall Apart. New York: Bloom's Literary Criticism. 2010. 177-96.

Tamuli, Girin & R.N. Koley. 'Yeshe Dorjee Thongchi' in *Silifor Review, Vol. II, No. II.* 2006. pp 41 - 48.

Tara, Tob Tarin. *Nyishi World.* Itanagar: Eastern Horizon. 2005.

Thakur, Amarendra Kumar. *Slavery in Arunachal Pradesh.* New Delhi: Mittal Publications. 2003.

Thakur, P. (ed.) *India's North East.* Delhi: Prakash Publication House. 1982.

Trevelyan, GM. *English Social History.* Bombay: Orient Longman. 1994.

Veeser, Aram H. (ed.) *The New Historicism.* London: Routledge. 1989.

Vidyarthi, L.P. & B.K. Rai. *The Tribal Culture of India.* New Delhi: Concept Publishing Company. 1976.

Waugh, Patricia. (ed.). *Literary Theory and Criticism.* New Delhi: OUP. 2006.

Wayne, Don. E. 'New Historicism' in *Encyclopedia of Literature and Criticism.* (ed.) Martin Coyle, Peter Garside, Malcolm Kelsall & John Peck. London: Routledge. 1991. pp. 791-805.

Weimann, R. *Literary History: Studies in the History and Theory of Historical Criticism.* London: Laurence and Wizhart. 1977.

West, Paul. *The Modern Novel.* London: Hutchinson. 1958.

White, Hayden. 'New Historicism: A Comment' in *The New Historicism.* (ed.) Aram H. Veeser. London: Routledge. 1989. pp 293 - 302.

Xaxa, Virginious. 'Tribes in India' in *Oxford India Companion to Sociology and Social Anthropology.* (ed.) Veena Das. New Delhi: OUP. 2003. pp 373-404.

Yapa, H. 1999: 'Tribal Village Council of A.P: Apatani' in *Tribal Village Councils of Arunachal Pradesh.* (ed.). B.B Pandey, D.K. Duarah, & N. Sarkar. Itanagar: Directorate of Research, Government of Arunachal Pradesh. 1999. pp 111-119.

Literary Sources (Assamese):

Barua, Prahlad. *Upanyas* (*Novel*). Dibrugarh: Banalata. 1990.

Bharali, Sailen. *Upanyas Bisar aru Bisleson* (*Judgement and Analysis of the Novel*). Guwahati: Bani Prakash. 1989.

Bharali, Sailen. *Adhunik Asomiya Sahityar Itihaas* (*The History of Modern Assamese Literature*). Guwahati: Chandra Prakash. 1992.

Bhattacharjee, Birendra Kumar. *Mopin Utsav* (*Mopin Festival*). Jorhat: Assam Sahitya Sabha. 1974.

Bhattacharjee, Pramodchandra (ed.) *Asomor Janajati* (*Tribals of Assam*). Jorhat: Assam Sahitya Sabha. 1962.

Bora, Dilip. 'Exa Bacharar Axamiya Upanyaxot Janajatiya Jibonor Chitra' (The Picture of Tribal Life as Reflected in the Assamese Novels of Hundred Years) in *Exa Bacharar Axamiya Upanyax* (Assamese Novels of One Hundred Years). (ed.) Nagen Thakur. Jyoti Prakashan: Guwahati. 2000. pp. 122 - 141.

Borgohain, H (ed.) *Binsa Satabdir Asamiya Sahitya* (*Twentieth Century Assmese Literature*). Jorhat: Assam Sahitya Sabha. 1967.

Borpujari, Jitanjali. *Asomiya Upanyasat Janajatiya Jivan* (*Tribal Life in Assamese Novel*). Guwahati: Chandra Prakash, 1999.

——. 'Exa Bacharar Asamiya Upanyaxor Samridhit Janajatiyo Lekhokor Abodan' (The Contribution of Tribal Writers in the Devolopment of the Assamese Novels of Hundred Years) in *Exa Bacharar Axamiya Upanyax* (Assamese Novels of One Hundred Years). (ed.) Nagen Thakur. Jyoti Prakashan: Guwahati. 2000. pp. 109 - 121.

Devi, Meera. 'Axamiya Upanyaxot Naribadi Chinta' (Feminism in the Assamese Novel) in *Exa Bacharar Axamiya Upanyax* (Assamese Novels of One Hundred Years). (ed.) Nagen Thakur. Jyoti Prakashan: Guwahati. 2000. pp. 273 - 297.

Neog, Maheswar. *Simantor Siksha aru Sanskritik Niti* (*Education and Cultural Policies in the Frontier*). Jorhat: Assam Sahitya Sabha. 1974.

——. *Asomiya Sahityar Ruprekha* (*An Outline of Assamese Literature*). Guwahati: Chandra Prakash. 1986.

——. *Adhunik Asomiya Sahitya* (*Modern Assamese Literature*). Guwahati: Chandra Prakash. 1987.

Sharma, G.P. *Upanyas aru Asomiya Upanyas* (*Novel and Assamese Novel*). Guwahati: Students' Store. 1995.

Sharma, Satyendra Nath. *Asomiya Sahityar Itibritya* (*An Account of the Assamese Literature of the Past*). Guwahati: Bani Mandir. 1963.

——. *Asomiya Upanyasar Gatidhara* (*The Current of the Assamese Novel*) Guwahati: Bani Mandir. 1976.

Thakur, Nagen. (ed.) *Exa Bacharar Axamiya Upanyax* (Assamese Novels of One Hundred Years). Jyoti Prakashan: Guwahati. 2000.

Magazines, Newspapers, Bio-data, and Reports (English and Assamese)

Bargohain, Homen. (ed.) *Amar Asom*. Guwahati. 6th April 2002.

——. (ed.) *Satsori*. Guwahati. 1st to 15th March 2006.

Bharali, Udayaditya. 'Aru Ekhon Saku Bhangil' (One More Bridge Has Broken) in Haider Hussain (ed.) *Asamiya Pratidin*. Guwahati. 10th April 2002.

Dai, Lummer. Inaugural Lecture in the 66th Annual Meeting of Asom Sahitya Sabha. Dibrugarh. 9th February 2001.

Dai, Nanni. *Biodata of Late Lummer Dai*. Naharlagun. 2003.

Dutta, Ajanta. 'Lummer Dair Sandhanat' (In Search of Lummer Dai) in Ajit Kumar Bhuyan (ed.) *Aaji*. Guwahati. 3rd February 2002

Gohain, Chandra Barpatra. 'Lummer Dai' in Ajit Kumar Bhuyan. (ed.) *Aaji*. Guwahati. 27th April 2002.

Government of Arunachal Pradesh and Rajiv Gandhi University. *Arunachal Pradesh Human Development Report 2005*. Government of Arunachal Pradesh, Itanagar. 2006.

Nath, Pankaj. 'Asamiya Buddhijibi, Rajnitikor Babei Bor Asomkhon Bhag Bhag Hol: Lummer Dai' (Lummer Dai: It was because of Assamese Intellectuals and Politicians that Greater Assam has been Fragmented) Anil Barua (ed.) *Dainik Asom*. Guwahati. 22nd March 1999.

Thongchi, Yeshe Dorjee. 'Mor Sahitya Sarchar Bisoye Mor Baktabya' (My Views Regarding My Writings) in Homen Bargohain (ed.) *Satsori*. Guwahati. 1st to 15th March 2006.

Index

Repu